Economics and resources policy

Economics and resources policy

Edited by
J. A. Butlin

Longman London

Longman Group Limited

Longman House
Burnt Mill, Harlow, Essex, UK
Published in the United States of America by
Westview Press, Inc., Boulder, Colorado

© Longman Group Limited 1981

First published 1981

British Library Cataloguing in Publication Data
Economics and resources policy.
  1. Environmental policy
  I. Butlin, J. A.
  301.31   HC79.E5   80-40247
ISBN 0-582-45074-8

Printed and bound in Great Britain by
William Clowes (Beccles) Limited, Beccles and London

Contents

List of contributors vii
Editor's preface viii
Acknowledgements x

Part One. The basic economics of the environment and natural resources

Introduction 2

Chapter 1 Environmental quality and resource use under laissez-faire. *J. A. Butlin* 3

Chapter 2 Economic policy and the threat of doom. *J. E. Meade* 9

Chapter 3 Natural resource economics: the basic analytical principles. *J. P. McInerney* 30

Part Two. Intertemporal and intergenerational problems

Introduction 60

Chapter 4 Economics and resources. *G. M. Heal* 62

Chapter 5 Economics of a throwaway society. *T. Page* 74

Chapter 6 Social welfare and exhaustible resources. *P. Grout* 88

Part Three. The economics of self-reproducible resources management: the sea fishery

Introduction 112

Chapter 7 Rational resource management and institutional constraints: the case of the fishery. *P. Copes* 113

Chapter 8 The economics of fishing; an introduction. *G. R. Munro* 129

Part Four. Economic incentives and environmental management

Introduction 142

Chapter 9 The contribution of economic incentives to solid waste management. *J. A. Butlin* 144

Chapter 10 An economist's view of pollution charges as regulatory instruments. *J. M. Marquand* 153

Part Five. International aspects of environmental management

Introduction   162

Chapter 11   A survey of international economic
             repercussions of environmental policy.
             I. Walter   163

Chapter 12   The role of international organisations in the
             control of transfrontier natural resources and
             environmental issues.   *M. Potier*   183

Index   201

List of contributors _____

John Butlin is a lecturer in the Department of Agricultural Economics, University of Manchester, England.

Parzival Copes is a Professor in the Department of Economics, Simon Fraser University, Burnaby, British Columbia, Canada.

Paul Grout is a lecturer in the Department of Mathematical Economics, University of Birmingham, England.

Geoffrey Heal is Professor of Economics at the University of Sussex, Brighton, England.

John McInerney is Professor of Agricultural Economics, University of Reading, England.

Judith Marquand is a Senior Economic Adviser to the United Kingdom Department of Industry.

James Meade is an Honorary Fellow of Christ's College, Cambridge, having been Professor of Political Economy at the College.

Gordon Munro is a Professor in the Department of Economics, University of British Columbia, Vancouver, Canada.

Talbot Page is a Senior Visiting Fellow, Division of Humanities and Social Sciences, California Institute of Technology, Pasadena, California, USA.

Michel Potier is Head of the Industry and Environment Division in the Environment Directorate of the Organisation for Economic Co-operation and Development (OECD), Paris, France.

Ingo Walter is Associate Dean, New York Graduate School of Business Administration, New York, USA.

Editor's preface

The student who, during his university career, takes an interest in the social aspects of environmental problems, is likely to find himself frustrated by one or more of a number of issues. He may find that the issues confronting him are unlike those that have been encountered in more traditional courses: for economists, measurement seems to rely less on traditional econometrics, and more on implicit valuation, and simulation methods. It is a situation in which externalities, rather than being fringe considerations to be acknowledged and then dismissed, are at the very heart of the problems under consideration.

The student who is concerned about the environment, and natural resource problems, may find that his emotive or instinctive feelings about these issues appear to conflict with the more objective and more detached view implied by economics. Pollution is undesirable, and yet society may be advised to tolerate positive levels of pollution. A sufficiently adverse deterioration in environmental quality could have a catastrophic effect upon western, and probably global, civilisation, and yet there appear to be trades-off between issues of environmental quality and those relating to the rate of economic growth, the balance of payments, the level of employment, and the rate of inflation.

The purpose of this collection of readings is to aid the student taking a course in environmental economics to place the issues in perspective. The text is designed for an undergraduate audience, and those readings that have appeared elsewhere have, with the permission of the holders of the copyright, been suitably abridged for this purpose. The book is designed to be used in conjunction with a conventional text on environmental economics or as an adjunct to a comprehensive series of lectures in environmental and natural resource economics.

The orientation of the text needs elaboration. The policy relevance of welfare economics applied to environmental problems is often stated, but students frequently appear to find it difficult to see clearly the connection between economic theory and environmental management practice. It is to this end that this book of readings is directed: to encourage students to see the relevance of economics to the evolution of environmental policy.

The theme which runs through this book is that economics has an important role to play in an environmental and resources policy context. The readings are, therefore, arranged as a progression, as the title suggests, from theory to practice. Much of the book is devoted, essentially, to a set of partial equilibrium problems, but initially the emphasis is on the general equilibrium context of environmental and natural resources management problems.

The book is structured into five parts: the first considers the basic economic theory of environmental and natural resources management. The second part considers, in some detail, the problem of intergenerational equity: the third is concerned with the economics of fisheries management, both from a theoretical and institutional viewpoint. The fourth part considers two aspects of the pollution control problem: the taxing of pollution and the reduction at source of the pollutants emitted from industrial and domestic activities. The final part is devoted to international pollution problems, and finally discusses the role of the institutional organisation in environmental and resources management problems.

A final caution on the use of this book is in order. It is not a textbook in environmental economics: there are several books that already fulfil this need, including D. W. Pearce's *Environmental Economics* (Longman, 1976). This book is to accompany the courses in which such texts are the primary required reading. Nor does the book attempt to cover all the areas of environmental and natural resource economics. Inevitably, the topics covered are limited, representing a mix of concerns that the editor perceived. Also, whilst the book is directed towards undergraduates, instructors will need to assist with Grout's and Munro's contributions (in Parts Two and Three). Both, however, are included because of the current relevance of the topics they cover.

Acknowledgements

This book of readings has had a prolonged gestation. My acknowledgements must, therefore, also serve as apologies. Firstly, I thank the contributors who, with varying degrees of alacrity, responded to my call for papers, and who have waited patiently, without complaint, to see their contributions appear. Secondly, I wish to thank the two typists, Miss Judith Darnton and Miss Jennifer Vaughan, whose dexterity in coping with transcript revisions and corrections was exceeded only by their patience in coping with me.

This is also an appropriate place to thank two people who have encouraged me professionally along what has seemed, at times, an extraordinarily arduous course. They are Professor David Pearce, of Aberdeen University, and Professor Tony Scott, of the University of British Columbia. Only they know how much I owe to them for their encouragement over a number of years.

It is appropriate on these occasions to do penance in public for the loss to family life that the production of the text has caused. My family has long tolerated my absence from the hearth on professional grounds, and it is their general tolerance, rather than that attributable solely to this book, that I hereby warmly and gratefully acknowledge.

Part One

Dans ce meilleur des mondes possibles *
The basic economics of the environment and natural resources

*In this best of all possible worlds (*Voltaire, Candide*)

Introduction

The concern in this section is with theoretical aspects. The introduction serves simply as a bench-mark, to give direction, and as a point to which those who become lost can return to begin once more their journey through the text. It is not to be regarded as comprehensive, the basic welfare economics being covered in most textbooks on environmental economics (although not always with the same emphasis on the intertemporal nature of the problem).

Professor Meade's economic evaluation of the neo-Malthusian School is comprehensive. Of particular importance is the demonstration that it is not only natural constraints that could sound the death-knell for the human race: even in the absence of these constraints it is possible for too high a rate of current consumption to bring about the collapse of a civilisation (a possibility that was unforeseen by Meadows and Forrester). In addition, the thorough-going evaluation by Professor Meade of the contribution that an economist could have made to one of the contemporaneous modelling exercises serves as an environmental economist's apprentices' guide. The caution that, in questions of environmental and natural resources management, one ignores biological, geological or economic factors equally to one's peril, is a lesson that all must draw from this paper.

Professor McInerney's work focuses more on the application of Fisherian capital theory to the problems of environmental and natural resources management, with Scott's concept of user cost as applied to natural resources (Scott, 1967) (essentially an intertemporal opportunity cost to irreversible resource depletion), carefully woven in. Using a basic framework for intertemporal allocation Professor McInerney is able to bring together the intertemporal efficiency aspects of exhaustible and self-reproducible resources management problems.

Taken together, these readings reflect the general equilibrium and interdependent nature of environmental and resources management problems. The problems in question are not a set of independent problems, totally unrelated to the type and level of economic activity, but rather adjuncts of both. A wider realisation of this fact is essential to understanding the role of the economist in environmental and natural resources management problems.

References

Scott, A. D. (1967) 'The mine under conditions of certainty', in M. Gaffney (ed.) *Extractive Resources and Taxation*, University of Wisconsin Press.

Chapter 1

Environmental quality and resource use under laissez-faire   *J. A. Butlin*

We shall approach the problem of understanding the fundamental economic reasons for environmental concern in a free economy by examining the set of conditions that would have to prevail for society to be satisfied with the current balance between the use of 'the environment' as a waste disposal sump and as an amenity, and between the rate of extraction of known deposits of minerals, and the supplies remaining for future generations. Having established these conditions, and the institutional framework necessary for their existence, we will then be able to decide whether or not these conditions were met in the 'real world'. This approach is taken as it provides both a review of the material with which readers are expected to be familiar, and a reference source to which they can refer back, if necessary, whilst reading further in the book.

Optimal resource allocation: static welfare optimum

In this review of welfare economics in a general equilibrium framework, we need to spell out as clearly as possible the fundamental assumptions that underlie the analysis:
1. All inputs into and outputs from the production/distribution/consumption stream at any stage are fully and completely owned by individuals in society;
2. No producer or consumer is able to influence the prices paid for, or quantities purchased of any of the inputs or outputs in (1);
3. Any risk associated with production, distribution or consumption can be fully insured against using an actuarially fair premium.

To ensure that the reader realises fully the relevance of these three assumptions we shall explore them more fully. The first, which implies that everything is owned by somebody (and nothing is owned by everybody) means that any exchange that takes place in the economy must involve the complete and final transfer of rights over the input or output concerned from the seller to the purchaser. The importance of the existence and transfer of property rights in environmental economics cannot be overemphasised.

The second condition, requiring that no one in either input or output markets has any market power, or, in other words, that all markets are atomistically competitive, is clear. Its importance in a dynamic context must also be realised. Not only must individual economic agents within a generation not have any market power; it should also be the case that no generation can affect the prices paid for natural resources in future generations. This aspect of the second condition could more frequently be mentioned.

The third condition needs careful elaboration. It requires that all risks that may arise in the future, at any time, can be insured against. Insurability against all possible future contingencies is an exacting requirement. Coupled with the requirement that there exists a complete set of futures markets, the institutional feasibility of an optimal configuration of the economy and the environment fast disappears. When the requirements of no within – or between-generation market power, and full information to all economic agents (about the set of all possible future states of the world, together with the probability to be attached to each state), then the likelihood of even the simplest set of conditions for optimality being satisfied disappears, almost without trace. Why, then, should these conditions be traced out, in this text as in any other on applied economics, with such care? The reason, mainly attributable to Weber (Shils and Finch, 1949) is that such an 'ideal type' as perfect competition with full information and no uncertainty can be used as a standard of comparison with the 'real world', with the result that the comparison goes towards the construction of a theory of social policy in the particular area concerned, in this case environmental and natural resource policy.

In this ideal world situation, then, what criterion of optimality can be used? Basically, the criterion used is that of Pareto optimality, in which, given the preferences of all the individuals in a particular society, one state of the economy is to be preferred to another if at least one constituent is better off in one situation than another, and nobody is worse off. Wars of words have been fought because it is apparent in the 'real world' that no policy change worthy of the name would have a universally neutral or beneficial effect. Thus, should a Pareto-optimal situation be one in which nobody was worse off and at least some people were better off *after* the gainers had compensated the losers, or could a situation be regarded as Pareto optimal if such compensation were possible if not effected? These 'compensation schemes' are attributable respectively to Hicks (1939) and Kaldor (1939). (The latter has been shown by Scitovsky (1941) to contain a contradiction, such that each of two states of the economy could be shown to be preferred to the other. Each is really a reflection of the inadequacy of using an essentially efficiency-based criterion for optimality to assess essentially distributional aspects of a particular policy change. Better to change the criterion for optimality (to reflect the growing concern over the distributional aspects of policy changes, and the apparently shrinking concern over the efficiency aspects) than to tinker with the Pareto criterion, the distributional implications of which are weak, and inconclusive over quite broad changes in the distribution of income.

We must for the time being put these things aside. What is the implication for the configuration of the economy and the state of the environment in this (economically) perfect world?

We can best answer this by considering the characteristics of this economy. Consumers, because they know the existing menu of prices that prevails in the economy can plan their purchases so as to make themselves as content as possible. They do this with an income that they have made as large as possible by hiring out the services of the factors of production that they own at the going market rate. By hiring some inputs, and buying others, again at prevailing prices, each producer aims for a maximum surplus over all opportunity costs, or supernormal profit. In this economic nirvana there would be no unwanted build-up or run-down of stocks, no excess demand for or supply of any good, including environmental goods. The satisfaction that society gained from this configuration of output would simply be the sum of the satisfactions enjoyed by all the constituents, – no more, no less. Any attempt to change the prevailing income distribution would leave more people worse off that it made better off. People at that point in time would be quite content with the levels of production of goods and services produced, and they would be quite content with the level of environmental quality. They would rather tolerate the existing levels of atmospheric and water pollution, the levels of urban congestion, and the level of environmental amenity prevailing, than any other.

This is a snapshot of the ideal economy run by 'Pareto rules'. We have not yet concerned ourselves with growth rates (positive or negative) or the rate of investment in equipment, or in natural resources, or in the absorptive capacity of the environment. Even in this most perfect of all possible worlds, the economic dynamics present particular problems.

Dynamic welfare optimum

Within a dynamic context, we face two broad categories of problems: the first set are really institutional, and require us to ask about the institutional framework that would be required to achieve the inter-temporally Pareto-efficient distribution of natural resources extraction. As we noted above, the institutional requirements are for a set of perfectly functioning and perfectly competitve forward markets for all goods and services traded now or that might be traded at any point in time, and/or a set of full and perfect insurance markets. These, together with all the other assumptions discussed fully above, would ensure that the present value of the sum of the social welfare of all future generations was maximised. This dynamic, Pareto-optimal intertemporal distribution of social welfare would, however, be distributionally, and ethically, neutral. Any attempt to reduce the rate of economic growth in one generation so as to increase it in some succeeding generation would reduce the discounted sum of social welfare over all generations, and would be less desirable than the original policy. The survival of the human race is either ignored or taken for granted. If the optimal plan is for mankind to cease at some finite time in the future there is no side-constraint in the criterion that would have us search for an alternative growth path, with the apocalypse approaching infinity. Again, the fault does not lie with the criterion. It does its job, which is to point out the most economically efficient growth path for an economy under a perfect institutional framework. ('Perfect', that is, for the operation of the criterion: again, there are no ethical or humanitarian connotations to the term 'perfect' as used in economics.) Concern over the intertemporal distribution of social welfare, and over survival of the human race, leads us to the second, broader set of problems that can be gathered together under the heading of 'social justice'. Whereas the Pareto criterion would have us determine the appropriate rate at which future generations' welfare should be discounted, social justice would have us find some other criterion for comparing the welfare of different generations, most likely a criterion for 'fairness'. The Pareto criterion will present us with a growth plan for the economy that presents future generations with dwindling stocks of exhaustible natural resources, whose real prices are consistently rising at the rate of interest. The rate of reclamation and recycling will be sufficient to augment the supply of raw materials so as to ensure that no excess demands arise for any goods. Depending upon the preference of the current generation for the present over the future (the 'pure' rate of social time preference in the absence of uncertainty), the ambient quality of the environment may stay the same, or deteriorate. Whatever the outcome, the current generation is sure that this is the most preferred growth path for the economy. If this calls for extinction of certain fish-stocks, exhaustion of certain mineral deposits, a decline in environmental quality, and the eventual self-destruction of the human race, *nil desperandum*: in a Pareto world, all that matters is maximum economic efficiency.

Resources, environment and the 'real world'

I hold the world but as the world . . .;
A stage, where every man must play a part,
And mine a sad one

(Merchant of Venice, I. i.77)

Antonio's sentiments in addressing Gratiano reflect the sentiments of any environmental economist. His sad role is to show why all is not likely to be well with the quality of the environment and the rate of resource extraction, in this institutionally less-than-ideal world. We can best proceed by cataloguing the imperfections, and then noting the qualitative effects that these individually may have upon the rate of resource extraction and the level of environmental quality. This will then set the backdrop for the remainder of the book, which is about the problems of the environment looked at from an economic viewpoint, and the contributions that economics can make to the prescriptive policy measures that may be proposed to ameliorate the situation.

Perhaps the most pervasive 'market failure' in the field of natural resources and the environment comes from incomplete individual ownership of certain goods and services. These range from the so-called 'free goods' (alternatively common property resources, or open-access resources) such as common grazing grounds, non-sovereign airspace or sea fisheries. For some of these, regulation is determined by national or international authorities. This is the case for most sea fisheries today, and also for the use of airspace above most countries. In these cases, individual private ownership would not be institutionally feasible. Use of these, either for transport, or as an environmental disposal sump, or as a medium from which natural resources are extracted, requires continual, unrestricted access to contiguous zones of the sea or air. The transactions costs of negotiating rights of access with a multitude of atomistic owners would be prohibitive, as would be the costs of enforcing rights of ownership and individual sanctions against access.

In principle, governments may regulate the use of these free goods (more precisely environmental goods in excess supply historically). In practice, regulation has been ineffective, as in the cases of fishery regulation and, until recently in some countries, air and water pollution, together with solid waste disposal on land. The problem is persistently one of incomplete ownership (or incomplete exercise of ownership). The government may hold sovereign rights to the air above, the seas around, and the public places within a particular country. However, this does not ensure that every user of these goods is charged the full social costs for such use, nor, alternatively, that regulations are passed which effectively will restrict use. The transactions costs of charging full marginal social costs may be so high and the change in production patterns and income distribution so great, that the transition to a world where full social costs of all activities involving the use of these natural goods and services are charged would be too socially divisive. The alternative, regulation by central government, may require such a large proportion of the total national budget to implement effectively that it, too, would be politically unacceptable.

With imperfect ownership of environmental goods (or what, from the above discussion, amounts to the same thing, namely prohibitively high costs of effectively enforcing the rights of ownership) there is a precondition for excess use of the environment as a disposal medium, excessively high rates of extraction of natural resources from the environment and a gross national product composed of more 'pollution intensive' goods than would otherwise be the case. The Gross National Product may be so heavily dependent on the production of pollution intensive goods that the 'externalities' generated lower the level of GNP, due, for example, to deteriorating standards of health among workers, or congestion in transport facilities. The effects of imperfect ownership manifest themselves everywhere, from the threat of extinction to many species of animals, to inordinate quantities of litter in public places (for whose disposal the individual litterers concerned have paid a zero cost), to the temperature inversions in Los Angeles and Tokyo, to the concern over the quality of the water in Botany Bay, Australia, and in the Mediterranean.

We are now at the point at which we can move on to the other readings in this volume. It is apparent that we can have at least two substantive bases for concern over the bequest of environmental quality and natural resources that one generation will leave to the next under a free market economy. The first is that in a world where externalities abound, uncertainty is rife and concentrations of market power are the rule rather than the exception, the actual allocation of resources will not be such as produce an optimal output mix (i.e. that which would be produced under the ideal situation). In general the actual output mix will be more pollution intensive and more primary resource intensive than the ideal. It follows that the level of environmental quality will be lower than that which would have prevailed in our ideal world. Also, there will be fewer environmental amenity services (beauty spots, wilderness areas, and national parks, for example). The readings in the remainder of this book are designed to help the reader understand more clearly certain fundamental economic aspects of particular environmental and natural resource problems, and also to appreciate the economists' contribution to their solution. (We do not claim, in one discipline, to be able to solve all the social problems deriving from environmental degradation. We would, however, suggest that to ignore the economic aspects is to miss out on an important link in the process of devising comprehensive and satisfactory solutions to this problem.)

References:

Hicks, J. R. (1939) 'The foundations of welfare economics', *Economic Journal* **49,** pp.696–712.

Kaldor, N. (1939) 'Welfare propositions and interpersonal comparisons of utility', *Economic Journal* **49**, pp.549–60.

Scitovsky, T. (1941) 'A note on propositions in welfare economics', *Review of Economic Studies* **9**, pp.89–110.

Shils, E. A. and Finch, A. A.(1949) 'Objectivity in social science and social policy' in E. A. Shils and A. A. Finch (eds and trans.) *Methodology of the Social Sciences.* Free Press of Glencoe, New York.

Chapter 2

Economic policy and the threat of doom*
J. E. Meade

An economist with no training in the natural sciences cannot determine whether the threat of doom is an immediate and real one. The answer to this fundamental question depends upon scientific and technological assessment of such matters as the ultimate effects of certain ecological and atmospheric disturbances, the technological prospects of substituting one material for another, and the prospects of a more direct harnessing of solar energy. There is much disagreement among highly qualified natural scientists on these questions. An economist will not grudge the natural scientists their little squabbles; but he would be foolish to try to judge between them.

That there should be considerable disagreement between the optimists and the pessimists among scientists and technologists is itself significant. The scientific and technological problems involved are very numerous; many of them are far-reaching and difficult of solution; and above all, the interrelationships between them are exceedingly complex. What will happen to human society over the next half century depends upon a very complicated network of feedback relationships between demographic developments, industrial and economic developments, technological developments, biological and ecological developments, and psychological, political and sociological developments. In each of the many subdivisions of each of these separate fields experts are confronted with difficult specific problems which they have yet to solve; but in addition to these specialised problems there remains the basic problem of how the developments in these various fields react upon each other. We need to see the system as a whole; and in our present intellectual atmosphere of expert specialisation it is precisely in such generalisation of interrelationships that we are weakest.

Methods of studying feedback relationships in dynamic economic systems as a whole and problems of decision-making in conditions of uncertainty are matters to which economists have devoted a great deal of thought in recent years. For this reason an economist can usefully address himself to these problems.

Nevertheless, in this paper only the basic economic interrelationships will be considered in discussing what we should do to meet the threat of doom.

*This was originally presented as the Galton Lecture at the Eugenics Society annual meeting, September 1973, and was published in B. Benjamin, P. R. Cox and J. Peel (eds) *Resources and Population*, Academic Press, 1973. It appears here with the permission of the editors.

That will be sufficient to give an idea of the principles involved in considering interrelationships in a dynamic social system, and it will certainly be enough for one lecture for one hour.

What then are the economic factors in the threat of doom? The work of Professors Forrester (1971) and Meadows (1972) is the most familiar. As the total world population grows and as economic development raises manufactured output per head of population, the total growth of economic activity will, they argue, press upon three different kinds of constraint: first, the limited land surface of the globe; second, the limited stocks of certain irreplaceable materials such as minerals and fossil fuels; and, third, the limited ability of the environment to absorb the polluting effects of economic activity.

These are indeed three basic economic limiting factors. To these three the economist would be inclined to add a fourth, namely the available supply of man-made assets – machines, buildings, and so on. Each of these limiting factors has its own distinctive features, which have important implications for the devising of economic policies.

The first group of limiting factors, typified by land, may be called 'maintainable natural resources'. Consider a farm of a given quantity and quality of land. There may be room on it only for one farmer and his family at a time, if a given standard of living is to be obtained from its cultivation. But when farmer A and his family have passed on, farmer B and his family can enjoy it. There may not be room for two at a time, but there is room for an unlimited number of families provided that they succeed each other in time.

The pressure at any one time of population upon the limited amount of land and its effect in reducing output per head because of the so-called law of diminishing returns is, of course, the limiting factor which has been so prominent in classical economic analysis since the days of Malthus and Ricardo. The fact that it is an old and familiar idea does not mean that it is a false or an unimportant idea. On the contrary, it is very relevant indeed in the modern world. In addition to this straightforward economic law of diminishing returns, there are other non-economic features of the pressure of population upon the limited amount of land space which may give rise to serious human problems – for example, the psychological ills which may result from too close crowding together. As this lecture will be confined to the more or less straightforward economic problems, the tendency as population grows for output per head to fall as the amount of land per head is reduced must stand proxy for all the evils resulting from a scarcity of maintainable natural resources.

The second group of limiting factors may be called 'non-maintainable natural resources'. Consider a stock of 1,000,000 tons of coal. Suppose that a family must consume ten tons of coal a year to maintain a decent standard of living and suppose that a family lives for fifty years. Then the coal stock will provide for the decent living of 2,000 families, no more and no less. Nor does it matter (provided the stock is already mined and available) whether these families exist all at the same time so that a human society of 2,000 families lasts for only fifty years, or whether these families all succeed each other so that a human society of one family lasts for 100,000 years. If ten tons of coal a year are essential for a family and if the stock of coal has a finite limit, then clearly any decent population policy through its birth control arrangements must plan, sooner or later, for the painless extinction of the human race; and, on the face of it, in so far as the supply of non-maintainable natural resources is concerned, it does not matter whether we have a large population for a short time or a small population for a long time. In this respect 'non-maintainable natural resources' are very different from 'maintainable natural resources'.

The distinction between maintainable and non-maintainable natural resources in the real world is not absolutely clear-cut. Land of a given quality may be maintainable if properly farmed; but it can also be mined if it is overworked or allowed to erode, so that its power to satisfy wants is, like that of a stock of coal, used up once and for all. On the other hand, by recycling the use of certain minerals an otherwise 'non-maintainable natural resource' may be capable of being used again and again provided that its users succeed each other in time.

Nor is the distinction between natural resources (whether maintainable or non-maintainable), on the one hand, and man-made capital resources such as machinery, buildings, etc., a clear-cut one. Land can be improved by, for example, a man-made drainage system; it is then a mixture of a maintainable natural resource and a man-made capital asset. Coal at the bottom of a mine must be brought to the surface by human action; when it lies as an available stock in the surface coal yard, it is a mixture of a non-maintainable natural resource and a man-made capital asset.

The possible scarcity of man-made capital assets constitutes a possible limiting factor which an economist would wish to add to the Forrester – Meadows catalogue. It is clearly not such a rigid limiting factor as the fixed and immutable supplies of natural resources. But it could nevertheless in certain conditions be the decisive factor. Consider a population which has a very low standard of living and is nevertheless growing rapidly. Because its standard is low it may be unable to save more than a very small proportion of its income without reducing its standards below the barest subsistence level. This may mean that its stock of man-made capital instruments can grow at only a very low rate, since all its productive resources, such as they are, must be used to produce goods and services for immediate consumption rather than to produce goods to add to – or even to maintain – the existing stock of capital equipment. If the population is growing at a high rate, capital equipment per head will be falling. At some point the availability of dwellings, schools, hospitals, tools, machinery, factories, etc. per head of the population may become so low that output per head falls below the bare subsistence level. There is a human crisis due to a lack of man-made instruments; and this crisis which might occur even though there was no shortage of natural resources and no pollution of the environment could be just as devastating as a crisis due to those other restraining factors.

The final limiting factor on economic growth is the pollution of the environment. Its study needs the co-operative work of economists, demographers, and other social scientists, with biologists, chemists, ecologists, and other natural scientists. However, only some of the main implications for the general principles of economic policy of the existence of these problems are considered here.

In the Forrester–Meadows models of the world community this problem is treated in the following way. It is assumed that industrial production pours out a stream of pollutants of one kind or another, the flow of which into a reservoir of pollutants, as it were, varies in proportion to the level of world industrial activity; it is assumed that the natural ecological and meteorological systems drain away and eliminate a flow of pollutants out of this reservoir, this outflow depending upon the amount of pollution in the reservoir. thus the winds disperse smog, the waterways cope with sewage, some wastes are degraded by bacterial action, and so on. Thus the degree of environmental pollution rises if the flow into the pollution reservoir from industrial and other economic activity exceeds the rate at which natural cleansing forces are evaporating the existing pool of pollution in the reservoir. But it is assumed that beyond a certain point the atmosphere becomes so polluted that the action of these natural cleansing forces is impeded. At this point the outflow of pollutants from the reservoir no longer increases as the stock of pollutants increases; on the contrary, as the flow of pollutants into the reservoir raises the level of pollution in the reservoir beyond this critical point the outflow is actually reduced; and there is then a crisis due to an explosive rise in environmental pollution which chokes economic and other human activity.

This may well be a good model of the existing relationships, although other than economic expertise is necessary to determine this.[1] But in any case without stretching the meaning of words too outrageously we may perhaps then talk of environmental pollution as causing shortages of certain environmental goods – for example, it causes a shortage of clean air, or a shortage of poison-free fish, and so on.

With these introductory remarks on the general nature of the four basic limitations to economic growth – namely, shortages of maintainable natural resources, of non-maintainable natural resources, of man-made capital assets, and of environmental goods – we can now present a much simplified model of the dynamic interrelationships between these factors. The model is in its essence of the kind constructed by Professors Forrester and Meadows modified in two respects. First, the model has been greatly simplified, in particular confining it to the economic relationships. Second, the structure of their model has been altered in ways which make it rather more congenial to an economist.[2] Criticisms and comments on the detail are not therefore criticisms of the Forrester–Meadows models.

Figure 2.1 shows a model of the production system. In the figures solid lines represent positive and broken lines negative relationships. Thus in Fig. 2.1 Output per head (O/N) is assumed to be higher: (i) the higher is the amount of maintainable natural resources (or Land) per head (L/N): (ii) the higher is the remaining stock of non-maintainable (or Exhaustible) natural resources per head (E/N); (iii) the higher is the amount of man-made Kapital assets per head (K/N); (iv) the more advanced is the state of Technological knowledge (T); and (v) the lower is the level of Pollution in the pollution reservoir (P). If N is the Number of persons in the population and L is the amount of Land, then land per head (L/N) is the greater (i) the greater is L and (ii) the smaller is N; and similarly for (E/N) and (K/N).

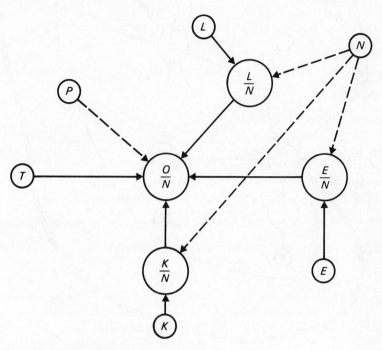

Fig. 2.1 The production system:
  N = number of persons in the population;
  L = amount of Land;
  E = amount of Exhaustible resources:
  K = amount of Kapital equipment;
  T = state of Technical knowledge;
  P = amount of environmental Pollution;
  O = total Output of goods and services;
  B = number of Births;
  D = number of Deaths.

In Fig. 2.2 the demographic relationships are added. $\triangle N$ represents the rate of increase in the total population and is the greater, (i) the greater is the total number of births (B); and (ii) the smaller is the total number of deaths (D). Three factors are assumed to affect the level of births and deaths: (i) Both total births (B) and total deaths (D) will be greater, the larger is the total population (N) subject to the forces of fertility and mortality. (ii) Births (B) will be reduced and deaths (D) increased by a rise in the level of environmental

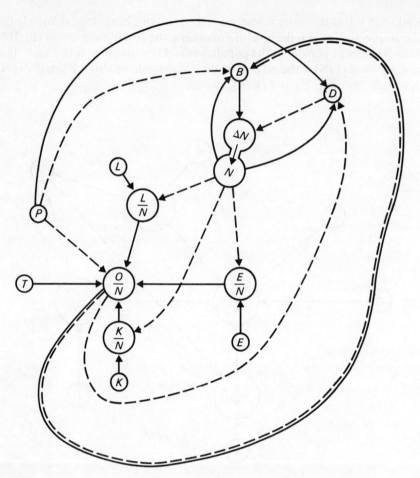

Fig. 2.2   The production system and the demographic relationships.

pollution (P); these are some of the links whereby a possible pollution crisis would show its effect. (iii) A rise in the standard of living (O/N) will reduce mortality (D). If standards are very low, a rise from the basic subsistence level is likely to raise fertility and births; but at higher levels of the standard of living, a further rise may cause a reduction in fertility; and accordingly in Fig. 2.2 (O/N) is joined to (B) both by a solid and by a broken line. But if the standard of living should fall very low, then the consequential fall in births and rise in deaths will show the links whereby a crisis for human society is caused by a production crisis (a low) (O/N) due itself to a high level of pollution (P) or to low levels of maintainable natural resources per head (L/N) of non-maintainable resources per head (E/N), or of man-made capital assets per head (K/N).

In Fig. 2.3 this simple model is completed by showing the links whereby the total level of economic activity may react in turn upon the availability of non-maintainable natural resources, upon the stock of man-made capital assets, and upon the level of environmental pollution. Total output (O) is the higher (i) the higher is the total population (N) and (ii) the higher is output per head (O/N). Non-maintainable natural resources are used up by the process of production and the rate of fall in the stock of such resources ($-\Delta E$) will, therefore, be the higher, the higher is the level of total output (O). The stock of man-made capital assets will be increased in so far as output is not consumed, but is used to invest in new capital instruments; and in so far as people save a given proportion of their real income, the rate at which the capital stock will rise (ΔK) will be the higher, the higher is the level of output and so of real income (O) from which savings can be made. But machines like human beings decay and die and so, just as a large human population (N) means that there will be many deaths (D), so a large stock of capital goods (K) will itself, through the need to replace old machines, reduce the net increase in the stock of machines resulting from any given level of newly produced machines (ΔK). Finally the rate of rise of the level of environmental pollution in the pollution reservoir (ΔP) will itself be greater, the greater is the level of total output (O). As has been already explained it is assumed that with a moderately low level of pollution the cleansing forces of nature will cause a flow out of the pollution reservoir which is greater, the greater the amount of pollution in the reservoir; but after a critical point is reached, the cleansing processes may become so choked that the flow out of the reservoir is reduced by a rise in the level of pollution in the reservoir. This double possibility is shown by a solid and a broken line joining (P) to (ΔP).

The interrelationships in Fig. 2.3 are already complex enough in spite of the great simplification of the reality which it represents. Indeed the complexities are certainly too great for it to be possible to generalise about the future course of events merely by inspection of Fig. 2.3. But in principle one should be able to tell the future course of all the variables in Fig. 2.3 if one knew three things:

First, the starting point, namely the present size of the population (N), the present size of the stock of non-maintainable resources (E), the present size of the stock of man-made capital assets (K), the present state of technical knowledge (T), the present state of environmental pollution (P), and the present availability of maintainable natural resources (L);

Second, the form and strength of each individual relationship shown by the arrowed lines of Fig. 2.3 – for example, the rate at which the stock of non-maintainable resources is depleted ($-\Delta E$) by the level of world output (0), and

Third, the future course of technical knowledge (T).

For since one thing is assumed to lead to another in a determinate way, if we know where we start, how each individual variable affects each other individual variable, and how any outside or exogenous variables like T will behave, we should in principle be able to forecast the future movements of all the variables for an indefinite future time. We can instruct a computer to do the donkey work for us, and thus forecast the future course of world developments.

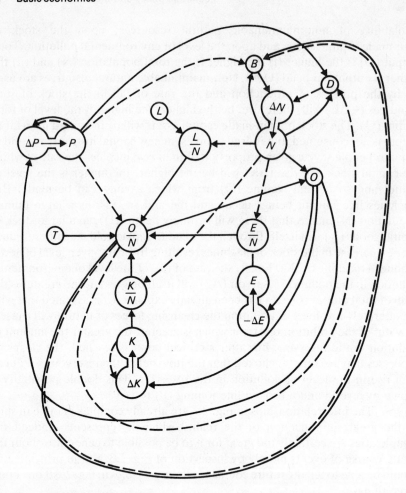

Fig. 2.3 The complete system.

One can next examine the effect of various changes in policy by telling the computer to assume that at some particular date there is some particular change in some particular relationship – for example, that as a result of a birth control campaign there is from 1980 onwards a reduction of a given amount on the influence which the size of the population (i.e. the number of potential mothers) has upon the number of births. One can then observe the effect on the future course of all the variables of this change in policy, after taking into account all the dynamic interrelationships in the system. And this is, of course, just the sort of thing which Professors Forrester and Meadows do with their dynamic models.

There are some basic truths which Professors Forrester and Meadows emphasise through their work.

First, there must be an end sooner or later to exponential growth of population and output, and the limit to such growth may come upon us unexpectedly unless we are careful. The facts about the present world demographic situation will be sufficiently familiar to illustrate the point. At present growth rates the world population doubles itself about every 30 years. If it were 3,500 million in 1970 it would be 7,000 million in 2000, 14,000 million in 2030, 28,000 million in 2060, and so on. Whatever the upper limit may be – and there obviously is *some* upper limit – we may hit it very suddenly. Indeed a mere 30 years before the final catastrophe we might be comforting ourselves with the thought that the world was after all only half full.

Second, the ultimate limit to growth may become effective either because of the exhaustion of non-maintainable natural resoures, or because of pressure upon the limited supply of maintainable natural resources, or because of the choking effects of excessive environmental pollution.

It is good that these basic points should be forcibly emphasised. But it is not necessary to construct a complex dynamic model for their demonstration. Clearly scarcities of natural resources and the choking effects of an ever-increasing reservoir of pollution would set ultimate limits to growth. An elaborate and sophisticated dynamic model is needed not to tell us this, but to tell us how soon and how suddenly the limits will be reached, which limit will operate first, how quick and severe will be the effects of reaching a limit, how effective a given change in policy will be in mitigating these effects, and so on. It is to answer questions of this kind that there is point in trying to construct models of dynamic interrelationships of the Forrester–Meadows kind.

What will happen with any set of dynamic causal interrelationships depends in a very important way not only upon the extent to which one variable (e.g. the standard of living) affects other variables (e.g. fertility and mortality), but also upon the speed with which the various influences operate. Indeed, one very real cause for concern about the present situation is the recent changes in the relative reaction speeds in different sectors of human activity. Many changes, and in particular technological changes, have speeded up very greatly. Disease and mortality have been reduced at unprecedented rates. New synthetic chemical and other materials, as well as new technological processes, have been introduced at previously unimagined rates. As a result world population and world industrial production are growing at speeds hitherto quite unknown.

But while some variables are changing in this way at much greater speeds than before, other reactions are just as slow as ever and in some cases have become more sluggish than before. Many of the new man-made chemicals and materials are slower to decompose than earlier natural substances and thus their effects (which in any case are novel and only partially understood) may be persistent and reach into the distant future.

To take another example, demographic reactions cannot be speeded up. It still takes a baby fifteen years or so of dependency before it starts to support itself, twenty years or so before it breeds, sixty years or so before it becomes an elderly dependant, and perhaps seventy years before it dies. Indeed, with the raising of school-leaving ages and with medical advances which keep people alive and active to greater ages than before, these demographic time-lags have in some respects been lengthened rather than shortened by present-day technological and social changes. Their importance may be illustrated in the following way. The continuation of high levels of fertility combined with relatively recent rapid reductions in the rates of child mortality have meant that there is an exceptionally high proportion of young children in many populations which are now growing rapidly. In these conditions population growth would continue for many years even if the fertility of women of child-bearing age were to be reduced instantaneously and without any delay whatsoever to levels which would merely replace the parents. For many years, as the present exceptionally high number of young girls grew up to motherhood, the total number of births would go on rising in spite of this immediate dramatic decline in the fertility of each individual woman. Such a population might well grow for another two or three generations and attain a size one-third greater than it was when the dramatic fall in fertility occurred.[3]

To take one more example, political delays between the observation of a change and legislative and administrative reaction to it remain as long as ever; and indeed the increasing insistence on democratic consensus in government may have lengthened the time needed to make acceptable a political decision which has obvious present disadvantages but whose future advantages are not at all obvious to the inexpert man in the street.

It is not therefore a sufficient answer to the prophets of Doom to say that their cry of Wolf has been equally relevant since the beginning of time. It has, of course, always been true that exponential growth cannot continue indefinitely. But what is unique about the present situation is the unprecedentedly rapid rate of population growth and of technological innovation (which represent exceptionally rapid approaches to the finite limiting ceilings) in a situation in which the results of population growth and of technological change are at least as prolonged and as persistent as ever and in which the ultimate policy reactions to danger signals are at least as slow as ever. Such time relationships do, of course, increase the possibilities of catastrophic overshooting of safe limits; and dynamic feedback models are in principle the proper instruments for assessing the importance of the relationship between different time lags.

One must therefore sympathise with attempts to think in terms of a dynamic model of these interrelationships; but an economist can only contemplate with an amazed awe the assurance with which Professors Forrester and Meadows provide answers to our anxious questions.

The real world is a hideously complicated system and it is inevitable that any dynamic model should be highly simplified. To be useful it must, on the one hand, be sufficiently simplified to be manageable by modern techniques of analysis and computation; but, on the other hand, it must not omit any of the structural relationships which may have a fundamental effect on the outcome, and the form and quantitative importance of the relationships which are included must be reasonably accurately estimated. Furthermore, the future course of certain outside, exogenous influences must be reasonably well predicted – a hazardous undertaking since it involves predicting the future effects of scientific and technological inventions without any precise foreknowledge of the inventions themselves; for if the inventions were already well understood, they would already have been made. These are very far-reaching requirements.

Economists have now much experience in coping with problems of this kind in searching for answers to much more limited questions. What is it which determines the demand for new motor cars? What is it which causes money wage rates to rise rapidly? What is it which governs businessmen's decisions to invest in new plant and machinery? And so on. But often, after the most detailed empirical enquiries, different hypotheses as to the structure of the causal relationships and as to the quantitative importance of any given factor in any assumed relationship provide conflicting results between which it is found difficult to choose even with the aid of the most refined statistical techniques. But the structure of the relationships and the numerical value of the parameters in a dynamic system can make a huge difference to the behaviour of the whole system; with one set of hypotheses the system may explode into a catastrophic breakdown and with another it may reach a stable equilibrium with or without moderate fluctuations on the way. But Professors Forrester and Meadows give results for an immensely complicated economic – social – demographic system of dynamic interrelationships for the whole world, having selected one assumed set of interrelationships and having used for each of those relationships estimates of the quantitative force of the various factors which in many cases are inevitably based on very limited empirical data.

For these reasons the conclusions drawn by Professors Forrester and Meadows are unquestionably surrounded with every kind of uncertainty. One must, therefore, ask what is the moral for present policy decisions if the future results of present policies are still extremely uncertain.

This question can be put in a very sharp form by considering one of the conclusions reached by Professor Forrester (1971) in his *World Dynamics*. He very rightly emphasises the fact that the ultimate effect of any given set of present policies depends upon dynamic interrelationships. Which influences work most quickly? To what extent are the evil effects of a given influence hidden at first and then operative with a cumulative, explosive effect? And so on. Professor Forrester concludes from his model that in order to prevent a worse ultimate disaster we should seriously consider the adoption of some very tough-line present policies:

> Instead of automatically attempting to cope with population growth national and international efforts to relieve the pressures of excess growth must be re-examined. Many such humanitarian impulses seem to be making matters worse in the long run. Rising pressures are necessary to hasten the day when population is stabilised. Pressures can be increased by reducing food production, reducing health services, and reducing industrialisation. Such reductions seem to have

only slight effect on the quality of life in the long run. The principal
effect will be in squeezing down and stopping runaway growth
(Forrester, 1971).

In other words we might be well advised to forget about family planning, to
discourage the green revolution in agriculture and the economic development
of undeveloped countries and to let poverty and undernourishment play their
role in restraining economic growth in the long-run interests of human welfare.

Professor Meadows dissociates himself from these startling recommendations for which Professor Forrester alone is responsible, but this
paradoxical conclusion of Professor Forrester is not necessarily nonsensical. It
could well be the correct prescription. But it depends upon a number of
assumptions built into Professor Forrester's dynamic model. It assumes that
while a successful birth control campaign may temporarily reduce population
growth and thereby raise living standards, it is not capable of preventing that
rise in living standards itself from causing a subsequent renewal of population
explosion. It assumes that economic development will be of a given polluting
character and that technology will not be capable of introducing sufficiently
non-polluting methods with sufficient speed. It assumes that the effect of
pollution is not a gradual effect, but stores up a cumulative reservoir of evil, as
it were, until there is a sudden explosive catastrophe. If these assumptions are
correct, then we ought perhaps to adopt tough-line present policies in order to
avert ultimate, total Doom. One should perhaps be prepared deliberately to
starve one person today to avert the starvation of ten people tomorrow.

But what if the outcome is uncertain? Should one starve one person
today to avert a 99 per cent probability of the starvation of ten persons
tomorrow? Perhaps, Yes; but should one do so to avert a 1 per cent probability
of the starvation of ten persons tomorrow? Pretty certainly, No.

Much work has been done in recent years, notably by economists, on
the pure theory of decision-making in conditions of uncertainty. In order to
make a precise calculation as to whether a given unpleasant decision today is or
is not worthwhile in view of its future potential benefits, one would in theory
have to have answers to the following set of five questions:

1.   What are the different possible future outcomes of today's decision?

2.   What probability should one assign to each of these possible outcomes?

3.   What is the valuation – or, in economist's jargon, the utility – which future
     citizens will attach to each of these outcomes?

4.   At what rate, if any, should we today discount the utilities of future
     generations?

5.   What valuation or disutility to us, the present generation, is to be
     attributed to the unpleasant policy decision which we are contemplating?

One could then calculate whether or not the disutility to the present generation
of today's unpleasant policy was greater or less than the discounted value of the

weighted average of the utilities to be attached to each of the possible future outcomes, each outcome being weighted by the probability of its occurrence.

The models of Professors Forrester and Meadows are intrinsically incapable of such treatment. They are deterministic and not stochastic in form, although in fact they are steeped in uncertainty. One could not in any case make precise calculations of the kind that have just been outlined about the present uncertain threats of Doom; but an appreciation of the principles of decision-making in conditions of uncertainty is helpful as a framework of ideas to inform one's hunches. Perhaps the disutility of Doom to future generations would be so great that, even if we give it a low probability and even if we discount future utilities at a high rate we would be wise to be very prudent indeed in our present actions. But we should not be prepared to carry prudence to the extent of abandoning our efforts to control present births and our efforts to raise agricultural outputs and the production of other essential products in the impoverished underdeveloped countries, though we should be prepared to carry prudence to the extent of a considerable shift of emphasis in the rich developed countries away from the use of resources for rapid growth in their material outputs towards the devotion of resources to the control of pollution, to the aid of the poor, to the promotion of technologies suitable for both developed and underdeveloped economies which save irreplaceable resources and avoid pollution, and to measures for the limitation of fertility.

We may conclude that the failure to deal with uncertainties is a serious weakness of the Forrester–Meadows type of model. A second serious weakness is its gross aggregation of many distinct variables. The model in Fig. 2.3 makes no distinction between events in different countries. It assumes only one output, making no distinction between different goods and services. It assumes only two uses of this single product, namely for personal consumption and for capital investment, allowing nothing for governmental uses for defence, space travel, supersonic aircraft, education, medicine, etc. It assumes only one form of pollution, making no distinctions between the pollution of air, water, or land or pollution by biodegradable wastes, by non-degradable wastes, by radio-active wastes, and so on.

This criticism is broadly true also of the models of Professors Forrester and Meadows, though they do both distinguish between agricultural and industrial production and Professor Meadows adds a third type, namely service industries. The introduction of this third distinction is also an important improvement. Services use up much less irreplaceable materials and cause much less pollution than does industrial production; and wealthy countries tend to spend a higher proportion of their incomes on services, thus providing a feature which mitigates somewhat the dangers of economic growth.

But all the models make no distinction between different countries, between different pollutants, or between different non-maintainable resources; and they make very little distinction between different products or different uses of products. This lack of disaggregation causes the models to exaggerate the threat of Doom in two important respects.

First, in so aggregated a model catastrophes are concentrated in their timing. Take the threat of a pollution crisis as an example. Let us accept the assumption that the evils of pollution often turn up unexpectedly with little forewarning when a reservoir of pollution rather suddenly reaches a critical level. In an aggregated model this must happen at the same time for every part of the world for every pollutant. *Ex hypothesi* remedial action is taken too late, and the result is, of course, catastrophic. But in fact atmospheric pollution in London rises unobserved to a crisis level in which smog kills a number of people; and belated action is taken to prevent that happening again; then the mercury danger reaches a critical level in a particular Japanese river; there is a local catastrophe; action is taken to deal with that; and so on. There is no intention to belittle these things or to deny that we should take these problems much more seriously than we have in the past. Nor, what is much more important, is there any wish to deny that there may be some much more far-reaching, global dangers which are creeping up on us, such as atmospheric changes which will turn the world into an ice-box or into a fiery furnace. Natural scientists should be given every opportunity and encouragement to speed up their efforts to decide which, if any, of such evils is threatened by which of our present activities. All that is being argued is that what in the real world might well take the form of a continuing series of local pollution disasters or of shortages of particular non-maintainable resources for which substitutes have not yet been found or of localised population control by a particular famine in a particular phase of development in a particular region are necessarily bound in an aggregated model to show up as a single collapse of the whole world system in a crisis of pollution, raw material exhaustion, or famine.

Second, an aggregated model cannot allow for substitutes between one thing and another, and some lines of economic activity use much more non-maintainable resources or produce much more pollution than do others. It may seem that too much of a consideration of only secondary importance is being given by stressing this lack of distinction in the models between different lines of production and different uses of products. Granted that there are some differences in the polluting effects and resource requirements of different lines of production and granted that economic growth may cause a shift in the relative importance of these different processes, yet, are not the shifts likely to cancel out to a large extent – some polluting processes gaining ground relatively and other polluting processes losing ground relatively? And, in any case, is not the net effect of relative shifts of economic structures of various industrial and other processes just as likely to be negative as it is to be positive on the balance sheets of pollution and resource requirements? In view of this, is it not perfectly legitimate to start with models which neglect such shifts?

An economist's immediate reaction is to point out that these models make no allowance for the operation of the price mechanism in causing one economic activity to be substituted for another. This point helps to explain why it is that economists are often less pessimistic than natural scientists in their attitude towards these problems. A large part of an economist's training revolves round the idea of a price mechanism in which that which is scarce goes up in price relatively to that which is plentiful with what in his jargon he calls 'substitution effects' both on the supply side and on the demand side. Producers will turn to the production of that which is profitable because its price has gone up away from the production of that which us unprofitable because its price has fallen, while consumers or other users will turn from the consumption or use of that which has become expensive to the consumption or use of that which has fallen in price.

In so far as a mechanism of this kind is at work it means that the changes of economic structure that are brought about in the process of economic growth will not be neutral in their effects on demand for scarce resources. They will be heavily biased in favour of activities which avoid the use of scarce resources and rely on the use of more plentiful resources. How far this process will help to put off the evil day depends, of course, upon the possibilities of substitution throughout the economic system; and it is here that economists are apt to be on the optimistic side. When a raw material becomes scarce and its price goes up, it becomes profitable to work ores with a lower mineral content, to spend money on exploration of new sources, to use scrap and recycling processes more extensively, to substitute another raw material, to turn to the production of alternative final products which do not contain this particular material, and – above all – to direct Research and Development expenditure towards finding new ways of promoting these various methods of substitution. Indeed this process of substitution permeates the whole economic system. Family budgets are sensitive to relative prices; in India where labour is cheap and capital goods expensive clothes are washed by human beings but in the United States where the reverse is the case this is done by washing machines. Agriculture is intensive in the Netherlands where land is scarce and expensive and is extensive on the prairies of Canada where it is plentiful and cheap. Business enterprises succeed by finding a new process which, at current costs of the various inputs, is cheaper and therefore more profitable. Moreover – and this is of quite fundamental importance – commercial research and development is expressly geared to find new processes which economise in scarce and expensive inputs and rely on cheaper and more plentiful inputs; and technology, as we all know, can be a very powerful factor in modern society.

This raises a particular problem. Is it any use fiddling with the price mechanism while the nuclear reactors burn? Are these considerations of any relevance at all to the great problems of environmental pollution which constitute the major threat of Doom? It is in fact precisely here that we need a major revolution in economic policy to make the price mechanism work. Environmental pollution is a case of what economists call 'external diseconomies'. When you drive out onto the streets of London you pay neither for the damage done by the poisonous fumes from your exhaust nor for the cost of the extra delays to other travellers due to the extra congestion which you cause. When you take your seat to fly your supersonic aeroplane over my house, you are not charged for the noise you make. When you treat your farmland with artificial fertiliser, you do not pay for the damage done to my neighbouring fresh water supply. When in your upstream factory you pour your effluent into the river, you do not pay for the damage to my downstream trout fishing. When you draw water for that extra unnecessary bath, you are not charged extra on your rates – unless you live in Malvern where domestic water supplies are metered and so charged and where the inhabitants seem to live a happy and clean life with an exceptionally low consumption of water per head. When you put out that extra dustbin of waste for municipal disposal, you are not charged extra on your rates. If you were you might not merely insist on your suppliers reducing the unnecessary packaging of the products which come into your house, but you might also collect your glass bottles and offer to pay their users to come and collect them for recycled use. The principle is the same for the most important and threatening examples of environmental pollution. We need politically to demand an extensive series of cost-benefit analyses of various economic activities and the imposition of taxes or levies of one kind or another at appropriate rates which correspond to the external diseconomies of these various activities. The price mechanism with its consequential process of substitution of what is cheap for what is costly could then play its part in the avoidance of environmental pollution just as it can in the economising of scarce natural resources. Business enterprise will be induced to avoid polluting processes. Technologists will be induced to steer their research and development into the discovery of new non-polluting methods of production.

Because of the magnitude of the problems that this issue raises, only one or two of the more general ones can be addressed here.

First and foremost there are the problems of deciding what are the probable ultimate results of different forms of economic activity. These are matters primarily for the natural scientist and the technologist. Will the global effects on the atmosphere turn the world into an ice-box or a fiery furnace? And what are the probabilities of these outcomes?

Second, there is the problem of evaluating the social nuisance caused by a given degree of pollution of a given kind. To make use of the classical example of a factory belching smoke, how does one measure in money values the cost of a given output of smoke when some people in the neighbourhood don't mind it much and others cannot abide it? Quite apart from the question how one adds up these different individual preferences, how does one discover them in the first place?

Third, a great deal of the damage done through environmental pollution is future damage. The use of DDT may confer important immediate benefits without any immediate indirect disastrous consequences; but it may be storing up great trouble for the next generation or the next generation but one. Quite apart from the technical difficulty in determining what will be the actual effects on the future of this pollutant, how does one evaluate that damage? How does one weigh the interests of future generations against the interests of the present generation?

Fourth, in most cases, if not in all, it is not a question of eliminating all pollution, but of keeping pollution down to its optimal level. The economist's favourite example, namely the smoking chimney, will be used. It may be prohibitively costly to eliminate all smoke, but not too costly to reduce significantly the output of smoke. To prohibit all smoke would leave the community without the smoke, but also perhaps without the product of the factory. To charge for the smoke the nuisance cost of the smoke might leave the community with some smoke nuisance, but also with the product of the factory. The latter situation might well be preferred. This is the basic reason for choosing, where possible, a policy of charging a levy or tax on the polluter which covers the social cost of the nuisance which he causes and then leaving him to decide how much pollution he will cause.

Fifth, in some cases – though these are much rarer than many administrators and technologists believe – it may be appropriate to act by a regulation rather than by a tax or charge on pollution. If the social damage is sufficiently grave, it may be wise to prohibit the activity entirely. Few people advocate discouraging murder by taxing it. But where it is possible to define and police a noxious activity for the purpose of regulating its amount, it is possible also to define and police it for the purpose of taxing it; and normally a tax on a noxious activity will be economically a much more efficient method of control than a direct regulation. Faced with a tax per unit of pollutant those who find it cheap to reduce the pollution will reduce it more than those who find it expensive to do so; and thus a given reduction in the total pollution can be obtained at a lower cost than if each polluter was forced by regulation to restrict his pollution by the same amount. Moreover, with a tax on pollution each polluter can employ the cheapest known method and, above all, will have every incentive to search for new and cheaper methods of pollution-abatement, whereas a direct regulation may well tie the polluter down to one particular method of abatement.

Sixth, in this use of fiscal incentives to avoid pollution, it is of great importance to tax that which is most noxious rather than to subsidise that which is less noxious. We all realise now that motor transport in large cities is causing intolerable congestion, noise, danger to life and limb, and atmospheric pollution. We all realise that private transport causes much more trouble per passenger-mile than does public transport. Both cause these troubles, but private transport causes more trouble than does public transport. The proper conclusion is to tax both forms of transport but to encourage the public relatively to private transport by taxing private transport much more heavily than public transport. The wrong conclusion is to leave the taxation of private transport where it is, but to subsidise public transport in order to attract passengers from the private to the public sector.

Such a mistaken policy has an additional obvious disadvantage. We already need heavy tax revenue to finance desirable public expenditures, and it is argued later that the new economic philosophy which we must evolve to meet the threat of Doom will make additional public expenditures necessary for such purposes as the redistribution of income in favour of the poorer sections of the community. The sensible thing to do now is to go round the whole economy taxing those activities which are noxious according to the degree of the social costs which they impose rather than starting to subsidise those competing activities which are somewhat less noxious. We can thereby help to kill two birds with one stone: we could discourage anti-social activities and at the same time raise revenue for the relief of poverty and for those other desirable public activities which we shall need to promote.

The points raised here have been confined to the use of taxes or other regulations to discourage economic activities which pollute the environment. In principle the same types of tax or regulation could be used to discourage economic processes which use up exhaustible materials; but the question as to whether it is necessary in this case to supplement the influence of the market price mechanism which will in any case raise the cost of scarce materials has not been addressed due to space limitations.

To summarise, it is a mistake to rely on models of future world events which assume a constant flow of pollution or a constant absorption of exhaustible materials per unit of output produced. Economic systems in the past have shown great flexibility. If we were to make the production of pollutants and the use of exhaustible materials really costly to those concerned, we might see dramatic changes. Indeed, there have already been some marked improvements in the cleansing of local atmospheres and waterways in those cases where the first steps of governmental action have been taken. There is no *a priori* reason for denying that if appropriate governmental action is taken to impose the social costs on those who cause the damage, there could be dramatic changes also in the more important and more threatening cases of the threat of Doom through pollution or through the exhaustion of resources.

Such is the first fundamental reorientation which we need in our economic policies, namely to set the stage by fiscal measures or by governmental regulation which will give a commercial incentive to free enterprise to select a structure of economic activities which avoids environmental pollution and the excessive use of exhaustible resources. But given the structural pattern of the economy, pollution and the exhaustion of natural resources will also be affected by the absolute level of total economic activity; and this means that there must be restraint over both the rate of growth of population and, at least in the developed countries, over the rate of growth of consumption per head.

This last consideration points to the need for a second fundamental change of emphasis in economic policies in the rich developed countries. Much modern competitive business seeks new profitable openings for business by commercial advertising which aims at generating new wants or at making consumers desire to discard an old model of a product in order to acquire a new model of what is basically the same product. Thus the desire for higher levels of consumption of unnecessary gadgets and of new models to replace existing equipment is stimulated at the expense of taking out the blessings of increased productivity in the form of increased leisure. The discouragement of commercial advertisement by means of heavy tax on such advertisement and the return to broadcasting systems which are not basically the organs for the stimulation of new wants by advertisement could be helpful moves in the right direction. Moreover, some steps could be taken to give incentives to producers to produce more durable products rather than objects expressly designed to need rapid replacement. For example, if cars were taxed much more heavily in the first years than in the later years of their life, consumers would demand cars which were durable and did not need rapid replacement. In general, if a heavy tax is laid on the purchase of a piece of equipment and if this discouragement to purchase is offset by a reduction in the rate of interest at which the funds needed to finance the purchase can be borrowed, there will be an incentive to go for durability in the equipment. Less frequent replacement will mean a lower tax bill, and at the same time the value of the equipment's yield in the more distant future will be discounted at a relatively low rate.

The need to set some restraints on the levels of total production suggests yet a third basic change of emphasis in our economic policies. If we wish to improve the lot of the poorest sections of humanity, then either we must rely on rapid and far-reaching growth of output per head or we must rely on the redistribution of income from the rich to the poor. In recent years both for the relief of domestic poverty and for the closing of the hideous gap between standards of living of the rich, developed countries and of the poor, under-developed countries the emphasis has been on economic growth. The extension of social services for the relief of poverty at home has, we have been told by our politicians, been impeded by the slow rate of growth of total output, it being assumed that any relief of poverty must come out of increased total production so that all classes may gain simultaneously. The raising of standards in the under-developed countries must, we have all assumed, come basically out of the growth of total world output, so that standards in the developed countries can be raised simultaneously with those in the under-developed countries.

This does not mean that we should avoid further economic growth. Indeed a rise in output per head, hopefully of a less noxious form than in recent years, is an essential ingredient in the relief of world poverty. A glance at the arithmetic of national incomes is sufficient to show that it cannot possibly be achieved simply by a redistribution of income from rich to poor countries. But it does mean that we would be wise to shift the emphasis significantly from a mere boosting of growth to a serious reliance on a more equal distribution of what we do produce, although we must face the fact that this inevitably mutliplies possibilities of conflict of interest between different classes in society.

But as soon as we emphasise redistribution we are faced with a very difficult dilemma. Anyone who studies the financial arithmetic of poverty is driven inevitably to the conclusion that if anything effective and manageable is to be done more help must be given to the large than to the small family. However one may do this, whether by higher family allowances or by more indirect and disguised means, it necessarily involves subsidising the production of children. If we aim at shifting our philosophy from a mad scramble for ever higher levels of production and consumption of goods, however unnecessary they may be, to a more humane and compassionate society in which basic needs are assured, if necessary at the expense of inessential luxuries, we come up against the thought that our children, who by the way never asked to be born, are also human beings with basic needs and that the more there are of them in a family the greater the total needs of that family if every member is to be given a proper start in life.

The same basic dilemma shows itself in a somewhat different form when we consider the closing of the gap between the rich and the poor countries. It is the poor countries with the highest rates of population growth which will be in the greatest need of foreign aid and technical assistance in order to undertake those projects of capital development (building new schools, new hospitals, new houses, new machines, new tools and so on) which are necessary simply in order to prevent a decline in the amount of capital equipment per head of the population. However disguised, does not this amount to the international subsidisation of those countries which are producing the most children?

Restraint on consumption per head is a means of restraining total demands on scarce resources which necessarily involves restraints on standards of living. On the contrary, restraint of population growth is a means of restraining total demands without any fall in standards of living. Population control may for this reason be put high on the order of priority for action to meet the threat of Doom though it raises a basic ethical question which cannot be discussed here. At what level is it legitimate to maintain standards for the born by denying existence at current standards to those whose births are prevented? It would appear that, however one might answer this basic ethical question, the population explosion is now such that restraints on fertility should constitute our first priority as a means for restraining the growth of total demands on scarce resources of land, materials, and environment.

In conclusion, it is necessary briefly to indicate one or two of the most important related issues. First and foremost, there is the distinction between the rich and the poor nations. The less developed countries fear that the concern of the richer countries with the quality of the environment–a luxury which the rich can well afford–will for various reasons impede economic growth in the less developed countries–a necessity which the poor cannot do without. Past experience has shown that a recession of economic activity in the Unites States and other developed countries has hit the under-developed countries by reducing the demand for their exports and by reducing the amount of capital funds available in the rich countries for investment in the poorer countries. Might not a planned restraint on the growth of real income in the rich countries have similar effects in reducing their demand for imports, their foreign aid, and the capital funds available for the development of the poorer countries and, indeed, in leading in general to an attitude unfavourable to industrialisation and growth in the poorer countries?

This fear must be exorcised. The stimulation of output per head in the poor countries is an absolute necessity for dealing with poverty in those countries. Such economic development is not incompatible with increased emphasis on population control, pollution control, and the recycling of materials. These things must not be confounded with policies to keep down the standards of living in the poorer countries.

A second set of major international problems arises from the fact that many of the problems which I have discussed cut across national frontiers. The supersonic aircraft of country A pollutes the atmosphere for country B. the whalers of country C reduce the catch for the whalers of country D. Country E may pollute a river, lake, or sea on which country F is also situated. Many of the controls which I have discussed will need international agreement and organisation..

And finally there is the problem of international disarmament. It is not merely that nuclear, chemical, and biological weapons of war would, if used, represent the ultimate pollution of the environment. There is a much more mundane day-to-day consideration. The production of armaments itself constitutes an appreciable proportion of industrial output in the developed countries; and it is concentrated on sectors of the economy which make heavy demands on material and environmental resources. Moreover, there is a very heavy concentration in the richer countries of governmental research and development on weapons of war which, if turned to such topics as the control of the environment, might transform the outlook. Disarmament could make a major contribution to our problem.

The development of the will and the institutions for international action in these three fields is essential for the successful moulding of any set of effective economic policies to meet the threat of Doom.

Notes

1. The model has been subject to extensive criticism. See A.S. Coles (ed.) *Thinking About the Future*, Sussex University Press, 1973, Ch. 7. (*Editor*.)
2. Economists will, however, notice that the model still has a number of glaring economic deficiencies. Quite apart from the need for greater disaggregation, discussed later, the production function does not allow for increasing returns to scale; technical progress is an entirely exogenous factor, responding neither to learning by doing nor to investment in research and development; there is no proper savings function; the standard of living is measured by output per head and not by consumption per head; inequalities in the distribution of income are ignored; investment is assumed to be maintained at a level sufficient to give full employment; and so on.
3. The demographic time-lags play no role in Professor Forrester's model, where no allowance is made for changes in the age-composition of the world population. But Professor Meadows introduced them into his *Limits to Growth*. The interested reader is referred to an Appendix to the original text on which this chapter is based.

References

Forrester, J. W. (1971) *World Dynamics* Wright-Allen Press Cambridge, Mass.
Meadows, D. L. *et al* (1972) *The Limits to Growth*, Earth Island, London and Universe Books, in New York.

Chapter 3

Natural resource economics: the basic analytical principles*   *John McInerney*

The field of natural resource economics

Probably everybody has some intuitive notion of what is meant by the term 'natural resources', but providing a rigorous and informative definition is not at all straightforward. At their simplest they might be defined as those economic factors in production or consumption which owe their origin and existence to natural phenomena, or to processes that occur autonomously in nature. This is an all-embracing statement which would seem to include anything that is neither man-made nor generated as a result of a manufacturing process that man has initiated – in short, everything that is neither 'capital' nor 'labour' in the classical economist's three-way classification of factors production. In this way all natural resources become equated with the traditional resource category of 'land', and this does capture one of their essential features – their initial availability precedes economic activity and is largely outside of man's influence. Natural resources are either God-given ('the original and indestructible properties of the soil') or originate from biological, chemical or geological processes that cannot at the present time be controlled at will.

On the other hand, such a definition is rather limiting if the aim is to structure a formal area of study called 'natural resource economics'; for it would appear that natural resources only remain so long as man takes no interest in them. Thus, coal lying in the seam would be seen as a natural resource, but once it has been hewn and transported to the pithead it becomes capital; a cod apparently ceases to be a natural resource when it gets tangled in a net, and a tree likewise once it is felled. Precise definition becomes progressively more difficult as human activity becomes more complex. If a natural lake is a natural resource, how would one classify the amenity services provided by damming a river? If wildlife is a natural resource, are not the animals protected in the Kenya game reserves–though still created by natural biological processes–somewhat 'unnatural' resources? Rainfall is one of natures's resources, but is it a capital resource if produced by cloud seeding? The list of examples in the grey area is endless, so obviously there is no clear division of resources into natural and others. But if we adopt a definition of natural resources that causes them to disappear as soon as they enter the conventional domain of economic activity, then we have succeeded only in

*This is a modified and corrected version of a paper which first appeared in the *Journal of Agricultural Economics* **27**, January 1976. It appears with the kind permission of the editors.

constructing one of those infamous empty economic boxes that litter the cupboards of our subject (Clapham, 1922). Treating oil deposits as a natural resource and unrefined crude as a capital good neither changes nor casts much light on the problems of world energy supplies!

If economics is to be functional as an applied discipline in this area, to serve as a guide to policy choices rather than as an exercise in intellectual tidiness, it needs to identify problem situations and group them into types which share some common thread in economic terms. The commonality between problems of land use, wildlife conservation, energy supplies and pollution is not meaningfully revealed by appealing to their connection with some natural processes which originated before man. It can, however, be found if we start from the idea of natural resources having a fixed initial availability – i.e. they exist as *stocks*. Clearly every resource exists as a given stock at a particular point in time, but a natural resource has one of two main distinguishing features. Either

1. The maximum stock of the resource that could be utilised is totally fixed, having been predetermined before man commenced any economic activity; or
2. To the extent that the available stock changes, it does so at a 'natural' biological or biochemical rate; this rate may not be constant over time, but biological or biochemical factors will prescribe a maximum rate of change (certainly with respect to increases in stock) that is outside of man's control.

The first characteristic is shared by resources such as the land area, metal ores, fossil fuels, scenic amenity, and other resources of a geophysical nature. The second characteristic is exemplified by forests, fish stocks, natural fauna and flora, fresh (as opposed to polluted) air and water supplies, and the other resources of a biological nature.[1] Human labour, too, might seem to fit into this second category, but from the economic (and social) point of view the problems of labour utilisation and natural resource utilisation are very different. It is to the economic aspects, therefore, that we must turn to finally delimit the natural resource area. Because of their stock characteristics, the essence of natural resources as a branch of applied economics is the central problem of *intertemporal allocation* – deciding how much of the existing stock of resource should be designated for use (consumption) now, and how much should be left *in situ* for the future. Although many other considerations may have to be introduced it is this focal problem, and the analytical techniques necessary to handle it, that provide the basis for 'natural resource economics' as a distinct area of study. We may therefore define this area as *the study of society's choices in the intertemporal allocation of resources (or resource services) derived from stocks which are either fixed or are changing at 'natural' rates*. This definition illustrates the three key features of the subject which serve to distinguish it from the study of other economic problems:

1. The problems are to be viewed at a societal as opposed to an individual level, and therefore attention is focused largely on social choice rather than the private choice that occupies so much of microeconomic theory.

2. Time considerations are of central importance to the analytical frameworks employed; there can be no entirely static theory of natural resource economics (although static aspects do enter in the form of the allocation of resources amongst competing uses at any one point in time – the bread and butter of traditional microeconomics).
3. The constraints on social choice are imposed by factors which are ultimately outside the control of man, for either the maximal resource stocks are immutably fixed, or change at rates which are not man-made.

From all this it is evident that the social choices to be made are fundamentally choices about the rate at which resource stocks are to be depleted or 'used up' – i.e. they are consumption choices.[2] Alternatively, since in choosing how much of a given stock to use one is by the same token deciding how much should be left untouched, we may equally view the situation as one of determining the optimal conservation policy. This emphasises that 'conservation' is controlled utilisation, a dynamic concept, as distinct from 'preservation' which is static.

It will be noted that our definition still leaves natural resources as an extended description of everything implied by the resource category 'land' in the traditional tripartite classification of productive factors into land, labour and capital–but refined so as to allow a more detailed economic treatment. 'Labour' is distinct because, although it is in some sense a resource which is subject to 'biological' rates of change, its supply of services is a flow rather than a depletion of the maximal stock and it is not a resource about which society as a whole can make arbitrary decisions as to its intertemporal allocation.[3] And 'capital' is distinct because, although it is a stock resource involving problems of intertemporal allocation, the size of this stock can be modified through time by industrial processes and at rates which society itself can freely determine.

The literature of natural resource economics

The study of natural resource economics is extremely wide and clearly of vital importance to a society attempting to consciously control its destiny. The availability of natural resources of all types is the underpinning to man's existence, and all production and consumption activity has always been a process of utilising natural resources. And yet surprisingly it is only relatively recently that the study of such resources has become established as a well-defined branch of economics – and even then it required a popular wave of concern over the long-run implications of a finite planet and the obvious degradation of our physical environment to alert economists in any number and persuade them to divert attention from their traditional fetishism with production, consumption and the pricing of consumed commodities. Apart from a few classic papers from earlier generations of economists (Gray, 1912; Hotelling, 1931), an identifiable literature on resource economics did not really make its appearance until the 1950s (Allen, 1955; Ciriacy-Wantrup, 1956; Gordon, 1957; Scott 1955) – and much of this was concerned with 'conservation' as a national target in an institutional setting rather than with the more general economic principles and analysis of intertemporal use. Even the formalised treatments of 'land economics' (the earliest established branch of resource economics) have taken a rather narrow and static view of the natural resource complex (Ely and Wehrwein, 1940). In the last decade the writings in this area have become many and varied, both for popular consumption and more esoteric interests, and offering projections for the future ranging from the scary to the sanguine. A feature of this literature has been an increasing focus on global concerns, with attention directed to such key issues as population growth (Ehrlich, *et al*, 1973), the finiteness of resource supplies (Meadows *et al*, 1972), the seemingly unavoidable paradoxes of resource ownership and access (Hardin, 1968; Dales, 1968), and pollution as a central feature of contemporary resource use patterns (Jarrett, 1966). As a consequence, a philosophy of 'environmentalism' now occupies an apparently respectable position in modern social thought (O'Riordan, 1976).

Despite all this attention, the fundamental principles for analysing natural resource use as a distinct class of economic problems have not been clearly enunciated. It is unfortunate, but perhaps inevitable, that the problems associated with the utilisation of natural resources tend ultimately to become very complicated, involving a web of technological, economic, social, institutional and finally political considerations. Even the economic aspects alone are not always straightforward. If treated in their entirety, the conceptual problems that must be faced (not to mention the practical ones) in specifying the utility-maximising allocation pattern of changing stocks between a multitude of competing uses over a continuous series of time periods soon become highly complex, the intertemporal aspect alone requiring resort to the fairly recondite mathematics of dynamics. This is reflected by those textbooks and articles which attempt to purvey the rigorous theory of natural resource use (e.g. Dasgupta and Heal, 1974; Herfindahl and Kneese, 1974; Solow, 1974) and which rapidly involve an appreciation of the calculus of variations, dynamic programming, and optimal control theory. For the uninitiated, or those who wish to gain but a general appreciation of the subject, this represents a formidable starting point–if not a complete barrier to entry. At the other extreme, the introductory treatments of resource problems tend to be at a relatively popular or descriptive level with little formal analytical economics applied to an integrated structure (Barnett and Morse, 1963; Jarrett, 1966); furthermore, they largely focus on the twin spectres of pollution and the environment as though these were the embodiment of all natural resource problems (e.g. Bain, 1973; Barclay and Seckler, 1972; Edel, 1973; Freeman *et al*, 1973), and seem to view the central economic issue primarily as one of externalities. Little appears to have been written between these two extremes which treats resource use as fundamentally an intertemporal allocation problem, yet does so simply enough to open up the area to, say, the second-year undergraduate student who wishes to relate the economics he has learned with the overall resource questions he reads about.

The aim of this chapter is to capture the analytical essentials of natural resource use problems while remaining within a framework of relatively basic economic principles. It makes no pretence at proposing constructive conclusions on the problems associated with the utilisation of any particular natural resource. Rather, it selects various components that are embodied in any introductory economics course and pieces them together in a particular way so as to cast some light upon typical resource use situations. Thus, it is claimed that using no more than the concepts of demand and social benefit, marginal cost of production and opportunity costs, discounting, and the ubiquitous equi-marginal principle, it is possible to cast resource use problems into an analytical framework which establishes both the type of social choice which must be faced, and the simple theoretical considerations that underlie an optimal allocation.

A framework for viewing natural resource use

In order to introduce the main ideas and enable the analysis to be conducted at this stage purely graphically, the following simplifications and assumptions have been adopted:

1. The available stock of any natural resource is to be utilised (allocated to consumption) over some finite planning horizon beyond which society has no particular interest – i.e. if necessary the resource stock can be totally depleted at the end of this time.[4] 'Society' can be interpreted as narrowly as a sub-national grouping whose prime source of income is a natural resource, or as widely as society on a world scale striving to ensure its continued survival on a finite planet.

2. The planning horizon for resource use is condensed into two discrete periods, each of unspecified length, which we may refer to as 'now' (t_0) and 'the future' (t_1). This may seem a gross simplification for a continuous sequence of time periods stretching into the future, but a planning horizon is by definition a finite period and therefore a truncated sequence; and a two-period representation of this sequence is quite sufficient to capture the essential multiperiod aspect of resource use decisions while avoiding unhelpful detail.

3. Society's preferences for the present and the future are assumed to be known and well defined, which means that we can identify the resource demand curves (or the marginal social benefits from resource consumption) for both t_0 and t_1. This may seem a rather brave assumption, but there is no escape from the fact that any attempt at a controlled use or conservation policy must inevitably involve information (or judgments) about the benefits to be derived from future consumption, so that they may be weighed against current consumption benefits. The difficulties with this assumption are therefore practical rather than conceptual.

4. Most natural resources have to be removed from the location where they were formed or grew before becoming of value to man, and therefore a 'cost of extraction' is incurred. In line with assumption (3) we shall take it that both current and future extraction costs can be determined – tantamount to assuming no unpredictable technological change in the resource extraction industries.

5. We shall abstract from the static problems of optimally allocating a given resource quantity between competing uses by assuming in general that each natural resource has effectively only one use. This is equivalent perhaps to assuming that the resource is demanded directly for consumption, whereas in reality most natural resources are subject to a derived demand which flows from the pattern of final goods consumed in society.

6. Finally we shall treat any particular natural resource as though it could be placed unambiguously into one of four distinct categories; these have some connection with the common division into 'renewable' and 'non-renewable' resources, but with important extensions as described later. Suffice it at this stage to characterise these four types as being non-renewable, recyclable, biological and flow resources; and as an *aide mémoire* the representative member of each type may be thought of as oil, steel, fish and land respectively. This four-way classification is paralleled by four basic economic models that encompass the range of choices society must make in the optimal intertemporal allocation of its resource stocks. It is an approximation, however, in that many natural resources are typically composites of characteristics and therefore do not obviously fit into just one of the four type classes.

Simplifying assumptions, such as those listed above, are such a well-known tactic in economic analysis that we tend to become immune to them. However, when applied to an area with which one is unfamiliar their crudity may suddenly appear to be glaringly unreal. The only defence offered here is that if initially a simple specification will allow an understanding of the basic situation and analytical technique to emerge, a more complete specification and treatment can always be built up subsequently by the usual extension into a multivariate mathematical model.[5]

The logic of intertemporal choice

The formal analytics of intertemporal choice were established by Fisher (1930) and can be found as part of the standard treatment in most modern microeconomic texts (e.g. Laidler, 1974; Nicholson, 1972) so that only a brief summary is called for here. One unit of a commodity consumed in period t_0 and one unit of the same physical commodity consumed at a later period t_1 are, from the economic standpoint, two entirely different commodities. Choosing how much of a resource stock to consume 'now' and how much to retain for 'the future' is therefore analogous to the constrained choice between alternative commodities so familiar in traditional static consumer demand theory. Notionally, at least, an indifference map can be drawn expressing the relative preferences between consumption in the two periods. The slope of an indifference curve (being the marginal rate of substitution of current for future consumption) will reflect the level of time preference – the rate at which future consumption will be sacrificed for current consumption. In particular, this slope when measured at the point of equal consumption in each period is given as $-(1 + \rho)$, is where ρ is the 'marginal rate of time preference proper'. It is because individuals are assumed, *ceteris paribus*, to prefer current to future consumption (i.e. they demonstrate a positive rate of time preference) that the value of future commodity flows are discounted to make them comparable with current-valued quantities. In natural resource use, where the decision-maker is a 'society' which may have longer planning horizons and manifest a concern for future generations, it is argued that decisions will be governed by lower rates of time preference than are demonstrated by individuals.[6]

Given this, a simple approach to the determination of an optimal natural resource use policy is shown in Fig. 3.1. When a fixed stock of resource (S) which neither physically appreciates nor depreciates over time, society can enjoy any intertemporal consumption pattern indicated by the constraint line SS, from total depletion in t_0 ('now') to total preservation of the stock for consumption in t_1 ('future'). By superimposing on this consumption possibility frontier the societal indifference map as between current and future consumption the optimal resource use and conservation policy is immediately indicated; C_0 units should be consumed now and $(S - C_0)$ units left unused to support a consumption level of C_1 in the future. If social preferences were reflected by the indifference curve in (a), for example, where the marginal rate of time preference proper is equal to zero $(1 + \rho = 1)$ the stock should be equally allocated between the two periods. The indifference curve in (b) reflects a positive rate of time preference $(1 + \rho > 1)$ leading to a high level of current depletion, while the indifference curve in (c) is indicative of a negative rate of time preference $(1 + \rho < 1)$ which would place extreme emphasis on resource preservation.

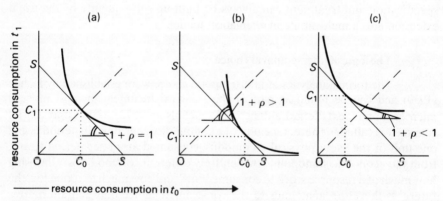

Fig. 3.1 Intertemporal allocation of fixed resource stock (S) under different levels of social time preference.

This approach to the identification of an optimal intertemporal resource use pattern[7] is illustrative in the very simplest cases of fixed non-renewable resource stocks. However, it is operationally restrictive in that it requires demand information in the less usual form of interperiod marginal rates of substitution (rather than the more typical demand projections), omits any consideration of the 'production' (extraction) costs of natural resource use, and cannot handle situations where stocks may change at an exogenously determined rate. More importantly, perhaps, it has nothing directly to say about any resource pricing policy that might be consistent with the optimal intertemporal allocation pattern. The problem that society faces is not just one of identifying the appropriate levels of current and future resource depletion, but also one of instituting mechanisms which will bring this about. No policy decision can fix or operate directly upon future consumption, for the future has yet to arrive. The achievement of the desired levels of future resource use, then, can only be influenced in a 'permissive' sense – by regulating current use so as to ensure sufficient (but not excess) retention for future needs. In other words, all that is needed and relevant from the operational policy standpoint is an analytical framework that will identify the appropriate resource price/ resource utilisation levels that should obtain in the current period. Conceptually there must exist some resource price (shadow price) which reflects the *real* cost to society of current consumption; if this price is established then the operation of resource policy might be left entirely to the market, adequate resource retention for the future being thereby brought about on its own without the need for physical controls on exploitation or other 'artificial' restriction of resource demand. This price may be determined by reference to the underlying resource demand and supply relationships – or, more precisely, the marginal social benefits and marginal social costs of resource consumption – and it is to this we now turn in search of a more fruitful framework for analysing the intertemporal choice problem.

Social benefits and costs and the elements of resource policy

Regardless of whether a natural resource is consumed directly or used as an input in the production of final consumer goods, one would expect it to be characterised by a 'normal' downward sloping demand curve; in other words, increased consumption in any period will confer progressively diminishing marginal social benefits. In our two-period setting we could therefore conceive of consumer demand and intertemporal preferences being based on two marginal social benefit curves, one for current consumption (MSB_0) and the other for future consumption (MSB_1). As we have already said, despite the severe practical difficulties that may be met in accurately specifying the latter, the construction of any forward-looking policy necessarily presupposes some knowledge of these future valuations; without them there can be no deliberate resource policy (other than total *laissez-faire*).

Although the demand side of the natural resource picture may be fairly straightforward, the supply side is less so. Since all natural resources exist as a stock at any given point in time it might appear at first sight that the supply curve must be perfectly inelastic, a vertical straight line rising from the quantity axis. However, the available stock merely specifies the maximal quantity that could be utilised, whereas the curve we require is the marginal cost to society of satisfying (supplying) the whole range of possible consumption levels in the given period. This is made up of two components. Firstly, although natural resources do not have to be 'produced' in the conventional sense, they do have to be harvested or extracted from their natural state, and this we may assume to be a process which requires the utilisation of other (capital and labour) resources under conditions of diminishing marginal productivity. Consequently, one component of resource supply in any period is a *marginal extraction cost* (MEC) which rises with increased resource consumption until the total stock is extracted (see Fig. 3.2).

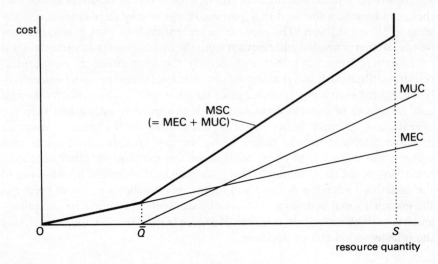

Fig. 3.2 The marginal costs of natural resource use.

Added to this is a highly important component of cost arising from the fact that because of the fixity of the resource stock, any unit consumed now is unavailable for consumption in the future. Current consumption therefore potentially comes at the expense of future consumption benefits, resulting in an opportunity cost or *user cost* being incurred. The notion of user cost was originally introduced by Keynes (1936) in connection with the utilisation of capital goods, but it is now of more central importance in the analytics of natural resource use (Nautiyal, 1970; Nicholson, 1972, Scott, 1953). In order to make clear its derivation, assume that a resource stock exists in excess of the maximum quantity that would be demanded in the future period. Initially, current utilisation (depletion) of this stock will not deprive society of future consumption benefits for these initial units are surplus to future requirements. But when resource extraction rises to a level (call it $\bar{Q}$ where the residual stock is just sufficient to satisfy the maximum future demand, then continued current consumption beyond this point will progressively rob society of benefits it would have received had it retained those resource units for consumption in the future. User cost is now a positive quantity, the user cost of each successive unit of current consumption being the (discounted) value of the future consumption that is thereby foregone. Under the assumption of a downward sloping marginal social benefit curve for future resource consumption, each successive unit of current consumption incurs a progressively higher loss of future benefits, yielding an upward sloping marginal user cost curve (MUC).

The real marginal cost to society of consuming a unit of resource in t_0 is therefore the sum of its extraction cost and its user cost. The summation of the marginal extraction cost and marginal user cost curves is shown in Fig. 3.2, giving the discontinuous marginal social cost curve (MSC).

We are now in a position to present the elements of an optimal resource use policy. According to the usual principles, society will gain the maximum net benefit from its resource stock when the marginal benefits and marginal costs of consumption are equal. As shown in Fig. 3.3, equating MSB with MSC implies consuming (depleting) Q^* units of the resource stock in the current period, thereby leaving $(S - Q^*)$ units untouched for the future; the appropriate (real) resource price that would bring this about on the market is P^*. The resulting intertemporal allocation is identical to the one that would be derived by employing indifference curve analysis within a two-period setting – and yet we have apparently managed to define it here solely in terms of current period quantities. The reason, of course, is that the introduction of user cost is really a simple means of bringing time considerations into the analysis; the inclusion of user costs as an added notional charge against current consumption causes an explicit recognition of its implications for the future, and thereby allows the true optimal level of current use to be identified within a single period framework.

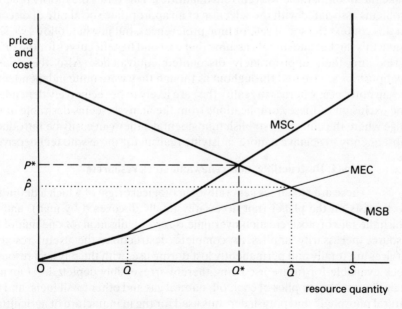

Fig. 3.3 Resource utilisation and pricing.

User cost can here be seen to be in the nature of an *externality* – a cost (loss of benefit) imposed on third parties (future generations) by the production/consumption activities of the current generation. By definition, externalities are not accounted for in market transactions, and it is because competitive resource pricing may tend to ignore the user cost element of natural resource exploitation that the need for intervention and direct conservation policies may arise. For it is conceivable (especially where property rights are not vested in the resource extractor, or where his planning horizon is shorter than that of society) that resource pricing and depletion might be determined on purely static considerations, with marginal benefits being equated only with those marginal costs that actually have to be met in the current period – i.e. MEC. This then leads to 'over-utilisation' and extravagant current consumption of the resource $(\hat{Q})$ encouraged by its underpricing in real terms $(\hat{P})$, with marginal social benefits from consumption being less than the true marginal cost to society.[8]

The foregoing discussion should have established the character, level and objectives of the simple analytics of natural resource economics, so that we may now broaden the treatment and examine the optimal utilisation policies for each of the four distinct types of natural resource mentioned earlier. Most of the simplifying assumptions have already been absorbed, but it will be (diagrammatically) convenient in what follows to assume that marginal extraction costs are constant within each of the two periods of the analysis. Furthermore, since an optimal intertemporal allocation of the resource stock is derived by relating marginal net social benefits[9] of current and future consumption, future costs and benefits must be discounted to a present value to make them comparable with current quantities. This raises the thorny practical problems associated with the selection of an appropriate social rate of discount (in this context the social rate of time preference), but in what follows we shall evade this thicket and merely assume that cost and benefit curves for the future period are their appropriately discounted equivalents. Also these latter relationships are treated throughout as though they were quite independent of the current ones, whereas in reality they are likely to be recursively dependent. The exclusion of these complications from the analysis seems desirable at this stage where the aim is to establish fundamentals; they can easily be introduced subsequently to achieve a more satisfying realism for those who feel deprived.

Type I: Destructible, non-renewable stock resources

These are resources generally of geological origin of which a definitive stock exists on the planet (not necessarily totally discovered by man), and for which the rate of stock creation over time is zero. 'Utilisation' of one unit of the resource necessarily implies its complete destruction; the useful resource services are totally and permanently lost during use, with the physical resource stock available for future use being thereby irreversibly depleted.[10] Obvious examples are the supplies of coal, oil, natural gas and other fossil fuels, and the natural phosphate and potash deposits used for the manufacture of agricultural fertilisers.

Determining an optimal utilisation policy is probably at its most straightforward with this resource type, involving simply the allocation of a fixed quantity between competing uses (here separated by time) according to appropriate marginal conditions – the kind of problem that economists cut their teeth on. In Fig. 3.4 the length of the horizontal axis OS is indicative of the magnitude of the resource stock; MSB_0 and MSB_1 represent the benefits derived from current and future consumption respectively, the latter having been drawn 'backwards' to accentuate the fact that future consumption is ultimately constrained by the amount of resource left unutilised after demand in t_0 has been satisfied. If these two curves are indicative of the level of resource demand in the two periods, MEC_0 and MEC_1 show the equivalent 'supply' curves in terms of the marginal costs of extracting the resource from its natural state.[11] In Fig. 3.4a the available stock exceeds the maximum that would be consumed over the planning horizon – the resource is to all intents and purposes available to society in unlimited supply (e.g. sea water for desalination or world coal stocks, the complete utilisation of which cannot be foreseen). Although theoretically user costs would be incurred at levels of current consumption in excess of $\hat{Q}_1$ (at which point the residual stock is just sufficient to satisfy the maximum future demand), they are of no relevance to resource use policy. Assuming rationality, utilisation in t_0 will not exceed $\hat{Q}_0$ where marginal benefit in consumption is equal to the marginal cost of resource extraction (i.e. net marginal social benefits from consumption are zero). Marginal social cost (MSC) is therefore identical to MEC_0 and no special market interference is necessary to modify resource price away from the level $\hat{P}$ which might be set competitively. Clearly, despite the fact that the resource stock is finite, active conservation policies are an irrelevant concern, for even with unrestricted exploitation an amount $(S - \hat{Q}_0)$ will be left which exceeds future demands.

(a)

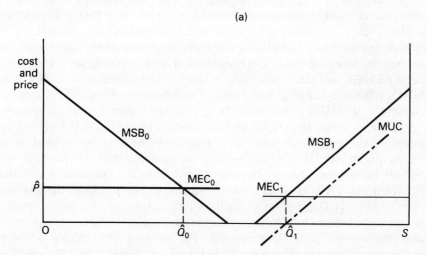

Fig. 3.4a Optimal intertemporal resource use where stocks are in surplus.

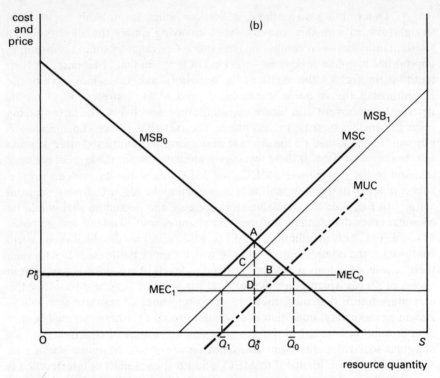

Fig. 3.4b Optimal intertemporal resource use where stocks are limiting.

The more realistic situation, where current consumption does involve a user cost, is depicted in Fig. 3.4b. Unrestricted exploitation in t_0 would lead to a residual stock $(S - \bar{Q}_0)$ being retained for future consumption, insufficient to satisfy the demand.[12] Resource utilisation beyond a level of $O\bar{Q}_1$ (where MSB_1 = MEC_1) deprives society of future net benefits, this loss being the user cost of current consumption. The MUC curve therefore measures the future marginal net benefits foregone by current consumption, and is thus the vertical distance between MSB_1 and MEC_1. As discussed earlier, the combination of MEC_0 and MUC_0 yields the marginal social cost of current consumption (MSC), and if equated with MSB_0 it will identify Q^*_0 as the optimal level of resource extraction to permit in t_0 resulting in an amount $(S - Q^*_0)$ being left for the future. The associated price P^*_0 measures the true cost of current consumption. Note that by partitioning the resource stock between the two periods in this way the net benefit derived from the last unit of resource extracted in t_0 (AB) is equal to the (discounted) net benefit of the last unit extracted in t_1 (CD), thus satisfying the equimarginal criterion for an optimal intertemporal allocation. Note further that as long as the full costs of current consumption are correctly measured, and (if necessary) policy action is implemented to ensure the resource is priced accordingly, then the needs of future generations should be exactly catered for.

Type II: Non-renewable stock resources with recyclable services

These again are typically geophysical resources of which a definitive and non-increasing stock exists. However, it now becomes important to distinguish between the physical resource as it occurs in nature and the consumption services that the resource provides. For in this case 'utilisation' of one unit of (physical) resource does not necessarily imply the total and permanent loss of future consumption; rather, the resource services are 'locked up' in a particular use for a particular period, from which they can subsequently be released by a process of industrial recycling for use again. There is no longer a simple one-to-one relationship between physical stocks of the natural resource and the stock of services that can be obtained from it. Iron, copper, lead, gold and the other metals found naturally in pure or ore form typify this group. Although initially the manufacture of a steel car body necessitates a permanent depletion in the stock of iron ore, the stock of steel *services* is only similarly depleted if the car body is irretrievably scrapped at the end of its useful life. The physical resource stock itself cannot be replaced, but the service stock can be – and at a technologically determined rate which is under man's control. In the limit, sufficient investment in the technological processes of recycling can totally recover the resource (though not in its original 'natural' form) and restore the stock of services to its original level. In this sense, despite continued consumption, theoretically society need never run out of the resource services unless it chooses to do so; with perfect recycling technology (impossible in practice, of course) all that is required is an initial physical resource stock sufficient to supply the services that are to become locked up in the immediate consumption cycle, with those same services being released to continually recirculate through time.[13]

Without this possibility of recovering already 'used' resource services by a process of industrial recycling, current utilisation would imply an equivalent future loss and the optimal extraction policy would be defined as previously for Type I resources. However, when recycling is feasible a unit of resource can be consumed in t_0 and then (its services) recaptured for consumption again in t_1 – so that current consumption need not necessarily deprive future generations. Recycling thereby provides a means of avoiding the user cost element in natural resource use – implying a lower real price and potentially permitting a more liberal current utilisation policy. However, recycling is not a costless process, requiring capital, labour and other resources in the recovery and reprocessing of the natural resource services, and so the complete re-utilisation of all units consumed in t_0 is not necessarily worthwhile. In general there will exist some optimal intertemporal pattern of consumption which is supported partly by depletion (i.e. use without recycling) and partly by recycled material, and it is this optimum that we wish to determine.

The exact nature of the possibilities afforded by recycling technology are an important factor in resolving this issue, particularly with respect to:
1. The proportion of the extant resource services that can be recaptured. For a variety of obvious reasons this will generally be less than one, though for

simplicity we shall assume in the following analysis that recycling implies the total recovery of resource services.

2. The length of time for which resource services are 'locked up' in the system and the rate at which they can be recycled. Again for simplicity we shall assume that recycling takes places at the beginning of t_1 on material that was utilised in the previous period. Among other things, this is tantamount to assuming that resource services are locked up for the whole of the period in which they were originally extracted, and that material can be recycled once only.[14] Society therefore faces a choice as to whether its consumption in t_1 should be supplied by depletion of any remaining natural stocks of the resource, or by recycling already extracted material.

3. The costs of the recycling process. Being an industrial process, it would seem valid to assume that the marginal cost of recycling (MRC) increases as more units for consumption are recovered in this way; and although recycling provides a means of avoiding the user costs associated with current resource consumption it may be so expensive as to make continued depletion the preferred alternative (as is the case at present with many 'scarce' resources). For purposes of analysis we shall assume that MRC rises less steeply than MUC.

As before we shall only concern ourselves with situations where resource stocks are an effective constraint on consumption, since situations where available stocks exceed the aggregate of current and future demands so that no user costs are incurred are less interesting. In general, resources can be extracted for current consumption quite freely up to the level $\bar{Q}_1$ where user costs would start to be incurred[15] (see Fig. 3.5). However, beyond this point

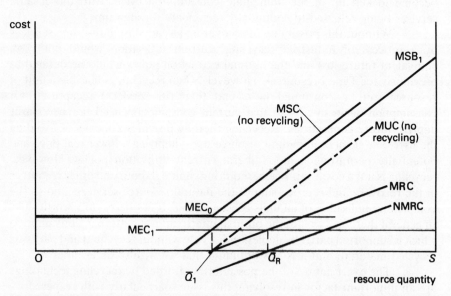

Fig. 3.5 Marginal costs associated with increasing consumption in t_0 given the possibility of recycling material for re-use in t_1.

the choices facing society require a little careful thought. If further resources were extracted and *not* recycled, this additional current consumption would result in a future loss of benefits according to MUC as discussed earlier. Alternatively if those resource units are later recycled there is no loss of future consumption benefits; the only 'loss' (i.e. charge against current consumption) is the cost of recycling them. Clearly, therefore, society faces a discrete choice as to whether any unit of current consumption should be supported by (a) simple resource depletion, and thus carry a user cost for the future consumption foregone, or (b) by technological recycling, and thus carry the costs of recovering those services for consumption again. And as long as recycling costs are less than user costs resource extraction can continue in t_0 without creating alarm (on condition that current MSB in consumption justifies it, of course).

Indeed, recycling carries with it a further advantage. For, compared to utilisation from the natural stock, any unit of resource service provided in t_1 from recycled material involves no extraction cost – this having already been incurred in t_0. The *net* cost to society of re-using a unit of resource services (NMRC) is the recycling cost incurred (MRC) minus the extractive cost (MEC$_1$) that would otherwise have been necessary. From Fig. 3.5 it is clear that NMRC does not become positive until extraction and utilisation in t_0 has reached $\bar{Q}_R$; by currently consuming resource units between $\bar{Q}_1$ and $\bar{Q}_R$ and

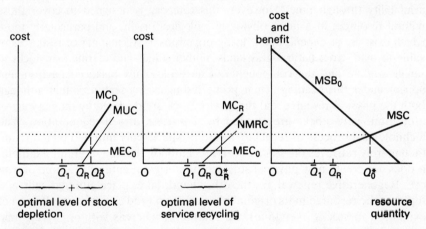

Fig. 3.6 The optimal consumption of resource in t_0 and its disposition between stock depletion and recycling.

then recycling them for use again in the next period, this aggregate consumption level is achieved more cheaply than would be possible if the resource services were obtained only from their 'natural' sources.

Since it will always pay to replace (recycle) resources up to $\bar{Q}_R$, user costs will not be met until current consumption exceeds this level. Continuing utilisation beyond this point should be distributed between depletive consumption (incurring a marginal cost of $MEC_0 + MUC = MC_{.D}$) and consumption-plus-recycling (incurring a marginal cost of $MEC_0 + NMRC = MC_R$) depending on which source involves the lower real cost for any particular resource unit. An amalgam derived by the horizontal addition of these two cost curves[16] gives the relevant overall marginal cost curve (MSC) of resource consumption in t_0, and the point where this equals the MSB of current consumption defines an optimal utilisation policy for the resource stock, as presented in Fig. 3.6. The current consumption level Q^*_0 will all be extracted in the current period, but an amount $(Q^*_R - \bar{Q}_1)$ will be recycled for future use while an amount $(\bar{Q}_1 + (Q^*_D - \bar{Q}_R))$ will be supported by simple depletion of the natural resource stock. The resulting overall level of consumption of resource services over the planning horizon is obviously higher than with the non-renewable Type I resources (since total consumption is equal to the complete stock plus recycled material) and therefore the real price to society is lower.

Type III: Destructible, renewable stock resources

This class shares some of the characteristics portrayed by each of the previous two types in that (a) 'utilisation' of a unit of the resource implies its destruction, a consequent depletion of the existing physical resource stock and the complete and irrecoverable loss of those resource services; and yet (b) the stock of resource services can be augmented again to enable a continuing availability through time. However, this category is designed to cover those natural resources of a more obviously biological origin, and particularly those which exist as, or depend upon, living organisms.[17] All manner of resources of value to man, from natural grasslands, timber stands and marine fish stocks to jungle wildlife and the components of an ecologically balanced environment appear under this heading. Their prime distinguishing feature is that although both the physical resource and the service stock are depleted by the very act of utilisation, new stocks are created by a process of self-regeneration. The technological process of recycling has to be specifically initiated by man, and then merely retrieves the stock of services embodied in the steel, for example; it does not create new physical stocks of the original natural resource of iron ore. Regenerative renewal, on the other hand, takes place automatically and replaces the resource in its original physical form (and can even result in a net expansion in stocks over time). Furthermore, whereas within a given time period the possibilities of renewal by recycling are dependent upon the amount of the original resource stock that has been utilised (the more steel 'in circulation', the greater the potential for recycling), with biological resources the extent of self-renewal is often more directly dependent upon the amount of the original stock remaining *unutilised*; the rate of biomass adjustment of a fish stock is strictly a function of that stock, for example (Tomkins and Butlin, 1975). This stems from the final key feature of this type of natural resource, which is that the renewal takes place at a 'natural' or biological rate which is largely outside the control of man, as opposed to the more widely controllable industrial rate at which technological recycling can take place. As long as the stock is not totally depleted (theoretically just one mating pair of blue whales or whooping cranes is sufficient) regrowth can occur to provide for future needs and thereby partly offset the user cost of current consumption, but there is some exogenously determined renewal rate r which remains the limiting factor.[18]

An optimal policy for the intertemporal allocation of such resource stocks must define the level of current utilisation that leaves a suffecent residual stock which, with additions due to natural regeneration, will just support future consumption and result in the appropriate balance between marginal net benefits in each period. This is one of the more difficult situations to explain and portray in solely graphical terms, and we shall find it convenient to develop the analysis through one of those four-quadrant diagrams so dear to the hearts of macroeconomists.[19]

Let us define r as the rate (or proportion) by which one unit of the natural resource, if unutilised in t_0, will have grown by the beginning of t_1. We shall continue to assume that the resources consumed in any period are harvested or extracted at the beginning of that period. Therefore, if Q_0 of the original stock of S units is utilised for current consumption there will be a stock equal to $(S - Q_0)(1 + r)$ available to support future consumption. From this it is immediately evident that the natural resource regeneration delays the onset of user costs; for even though current consumption may deplete the stock to a level insufficient to support future demands, this residual stock will have expanded by the time future consumption commences. Secondly, however, when user costs are finally incurred they are *higher* than in the previous cases (given the same MSB and MEC curves). The reason for this is that each unit of current consumption deprives the future generation of the consumption benefits of $(1 + r)$ units – i.e. the unit currently consumed plus the natural growth it would otherwise have undergone.[20] The optimal resource utilisation and conservation policy is therefore highly dependent upon the magnitude of r, thus making efforts directed nowadays towards increasing the rate of regeneration of biological resources an important element of policy.

The MSB and MEC curves for the current and future periods are shown in the north-east and south-west quadrants respectively of Fig. 3.7. The stock in the two periods is linked through the south-east quadrant by a line representing the growth of unutilised resource stock. The assumption of a constant regeneration rate yields a straight line of slope $(1+r)^{21}$; this indicates that, for example, if the total stock S is unutilised in t_0 then a stock level of S $(1+r)$ will be available for consumption in t_1. However, consumption in t_1 will never exceed $\hat{Q}_1$, this being the point where marginal net benefits from future consumption are zero ($\text{MSB}_1 = \text{MEC}_1$); a sufficient residual stock level to support this will be ensured if at least $(S - \bar{Q}_1)$ resource units are left unutilised in t_0, since by definition $(S - \bar{Q}_1)(1+r) = \hat{Q}_1$. Thus, if resource consumption in the current period were at $\bar{Q}_1$ or less there would be no adverse effects on future consumption, and no user costs would be incurred; but at utilisation levels beyond this an increasing user cost element would be associated with each additional unit consumed. As explained above, the MUC for each unit of current consumption is equal to the marginal net benefit foregone from $(1+r)$ units of future consumption, due to the combined loss of both the original resource unit plus its natural growth. Because of our assumption of constant marginal costs of extraction, if we drew the marginal net benefit curve for future consumption it would have the same slope as MSB_1. The MUC curve drawn in the north-east quadrant of Fig. 3.7 then has a slope $(1+r)^2$ times that of MSB_1.[22]

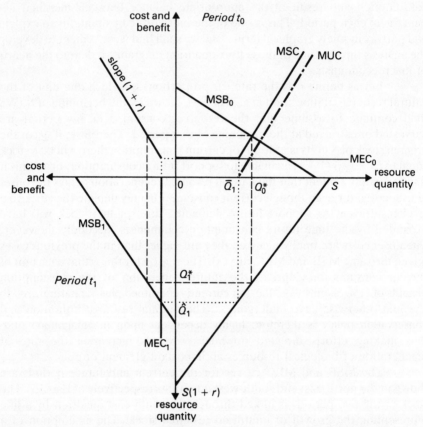

Fig. 3.7 Optimal intertemporal allocation of a biologically renewable resource stock.

When combined with MEC_0 this user cost curve gives MSC, the true cost of each unit consumed in t_0 which should guide current use decisions. The 'equilibrium' utilisation and conservation policy for t_0 is defined where MSC and MSB are equal for the current period. This results in a current utilisation of Q^*_0, with $(S - Q^*_0)$ being left to 'grow on' to support a future consumption level of Q^*_1.

By setting current price and resource use in this way, an intertemporal utilisation pattern is achieved from the resource stock which achieves the appropriate balance between current and discounted future consumption benefits. But it should be noted that this is *not* an equality between interperiod marginal net benefits. Here lies a crucial analytical point which flows specifically from the regenerative nature of the resources in this category. With both Types I and II, one unit of the physical resource (or service) can be carried forward into the future unchanged; so the benefits yielded by consuming a resource unit in t_0 can be quite simply compared with those that would be derived from consuming the same unit in t_1. However, a unit of Type III biological resource, if unconsumed in t_0, necessarily becomes $(1+r)$ units when carried forward into t_1. In determining an optimal intertemporal allocation, then, the valid comparison is between the benefits yielded by consuming any particular unit now and the benefit from consuming the $(1+r)$ units it would yield in the future. Consequently, an optimal resource use policy in this case is one where the net social benefit from the last unit consumed in t_0 is equal to the net benefit gained from the last $(1+r)$ units consumed in t_1–i.e. when $MNB_0 = (1+r) . MNB_1$.[23] The line in the north-west quadrant of Fig. 3.7 has a slope of $(1+r)$ and is drawn in such a way as to relate the interperiod net benefits in the appropriate manner, thus demonstrating the equilibrium through all four quadrants of the model.

Type IV: Non-destructible, non-renewable stock resources with renewable service flows

This final category, which is typified by that composite resource we refer to as 'land', becomes necessary because it and some other important natural resources do not fit easily into either of the previous three groups. The relationship between the physical stock of the resource and the stock of its productive services is quite different with land compared to the other resource types. The services naturally provided by a given area of land, whether they are housing, transport, amenity, food production services, or a multiple of these, can be obtained seemingly for evermore; despite the fact that the physical land stock is finite, the service stock would appear to be infinite – for regardless of the level of consumption of land services they can never be depleted. However, the utilisation of this stock of services within any given time period is not subject to such free choice as it is with resource stocks such as oil, iron ore, or fish. In principle the services provided by a unit of land are made available 'autonomously' at a constant rate over time, and this defines the maximal rate at which society may utilise the resource. 'Utilisation' of the land therefore strictly means only consuming its current flow of services, and involves no depletion of the physical resource stock and no diminution of its future service flow. Furthermore, it makes no difference whether the services are consumed or not; if they are not taken up those particular services are lost for ever and cannot be recaptured, the same as if they had been consumed – but either way an equivalent replacement set automatically becomes available in the next period. Although as a natural resource land exists as a physical stock, therefore, from the economic point of view it can only be viewed as a perpetual or constant flow resource, the size of the physical land stock determining the rate at which the total service flow becomes available. The infinite stock of services mentioned above only has relevance over an infinite time period; the same applies in principle to other 'inexhaustible' natural resources such as solar energy, rainfall, water resources, scenic amenity and the aesthetic services provided by our environment.

Despite the fact that there can be no meaningful static concept of the available stock of services when the resource is in actuality a fund of dated service flows, the intertemporal utilisation problem would seem to involve no great complexity. Take the simplest case of those natural resources which merely produce their service flow virtually independent of man's activity–for example, the sun giving off solar energy services, a natural beauty spot giving off visual amenity services, or agricultural land giving off food production services. The capture and current consumption of these services has no effect on future flows, nor can they be transferred backwards or forwards intertemporally, so a process of static constrained optimisation would seem quite sufficient.[24] With no user costs incurred, consumption in any period should continue until the current social marginal cost of obtaining those services (equivalent only to MEC) equals the current MSB, or until the available service flow is fully taken up, whichever occurs first. Resource use and pricing[25] can be determined independently in each period, and no forward-looking conservation policy is called for.

The complication and dangers of such a static approach arise with land use, and with various related resources such as amenity and 'the environment', because society can make choices in one period which may alter the *nature* of the service flows and thereby affect the consumption possibilities available for the future. Depending on the particular societal choice, certain natural resources can be used to produce different (and often mutually exclusive) *types* of service flows; land may be used for housing agricultural production, for example, and a canal for recreation or effluent disposal. It is because such decisions frequently infer virtually *irreversible* transformations in the nature of the service flows that we must reintroduce the intertemporal aspects into our analytical framework.

When viewing land use within the framework of intertemporal allocation, we need to recognise it as a resource possessing two distinct major characteristics, and therefore providing two quite different types of resource services.

1. Firstly, land provides a service which is simply *space*. This service flow is indestructible, unchangeable over time, and independent of the actual economic activity to which the land area is allocated.[26] One hectare of land, in whatever use it may from time to time find itself, will still always provide annually one hectare of space into perpetuity, and current use of these services can in no way affect their future availability.
2. Secondly, land provides a flow of production/consumption services depending on the use to which this space is allocated (farming, forestry, housing, transport, car parking, recreation, visual amenity, water

gathering, etc). This service flow is variable over time in type and quantity depending in large measure on man's activity, and therefore the intertemporal allocation of these services is of some consequence (and it is here that we come into contact with the conventional treatments of land use and land economics).

Within any given type of land use, the level of current consumption may cause depletion in the service flow for future periods. The mismanagement of soil resources leads to fertility depletion and erosion, for example; the non-maintenance of the housing stock, the overpopulation of amenity areas and the poulation of water courses are all current consumption activities which reduce the availability of the relevant services to future generations. However, these service flows can be re-augmented either naturally or technologically, and so the optimal intertemporal utilisation of the resource services can be treated within the frameworks proposed earlier for resource Types II and III. For example, effluent discharges into waterways (i.e. the rate of utilising the cleansing services of water) can be controlled at levels which allow the complete regeneration of water purity from one period to the next; or expenditures on house repairs and maintenance (i.e. investment in 'recycling' the consumed housing services) can be set at levels which result in the housing stock being maintained, or depleted through time only at the desired rate.

The more difficult intertemporal allocation decisions with respect to land involve *changes* in use, because many of these have implications of irreversibility. When land is transferred from agricultural to urban uses, for example, the service flows are affected in the following manner:
1. The flow of space services remains unaltered – as it does with any change of use,
2. The flow of agricultural production services is reduced to zero, and neither natural, biological, nor technological processes can regenerate this service flow,[27]
3. The lost agricultural land services are replaced by an equivalent flow of urban land services (housing, roads, etc).

Changes in land use do not, therefore, necessarily imply that anything is 'lost'–merely that the nature of the service flow is totally transformed. Inasmuch as this transformation is irreversible, the intertemporal allocation problem involves specifying that current pattern of land allocation between competing uses which will enable society to gain the maximum total benefit from both current and future consumption of its fixed land stock in all its different uses. In other words, any land use changes made in the current period must consider the gain or loss in future net benefits arising from the transformation of the service flows – what might be called the 'user cost' of the change.

Consider, for example, the optimal pattern of land use over two periods, where land is initially in agriculture and is regarded as having only one alternative use – housing. Figure 3.8 shows the current marginal net benefit[28] curves for agricultural land services and housing land services in relation to the fixed stock of land S. An optimal land use policy defined solely for the current period would be where the MNB in agriculture and housing uses are equal; this suggests that Q_H units of land would be transferred from agriculture (the prior use) into housing, leaving $(S - Q_H)$ units in agriculture. However, change of land use in this direction is technically irreversible, as land allocated to the provision of current housing services cannot be readjusted to provide future agricultural services. On the other hand, this constraint is undirectional, for land allocated to the provision of current agricultural services can provide either agricultural or housing services in the future, whichever is preferred. Care must be taken in the current period, therefore, to ensure that no transformation of land services is made which, although optimal for the current period, results in insufficient land being available for the provision of agricultural services in the future. In Fig. 3.8, the allocation of Q_H units to housing may not, in fact, be optimal if an area in excess of $(S - Q_H)$ would be required for agriculture in the future. Consumption of current housing can be seen to involve something in the nature of a user cost (measured in terms of future agricultural benefits foregone), and once such costs are incurred further land use changes must be made with caution if future benefits are not to be jeopardised.

Figure 3.9 shows both current and future marginal net benefits from housing and agricultural land services. If more than Q_1 units of land are transferred to housing in the current period, a suboptimal balance of agricultural and urban land will remain in the future. Further transfers out of agriculture in t_0 must therefore weigh the current additions to total net benefit against the (discounted) reduction in future total net benefits (the user cost of current housing, measured in terms of the agricultural production lost in t_1). If

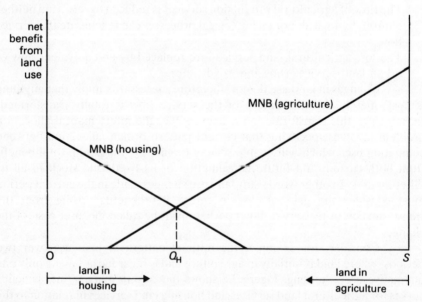

Fig. 3.8 Optimal allocation of land between competing uses, current period.

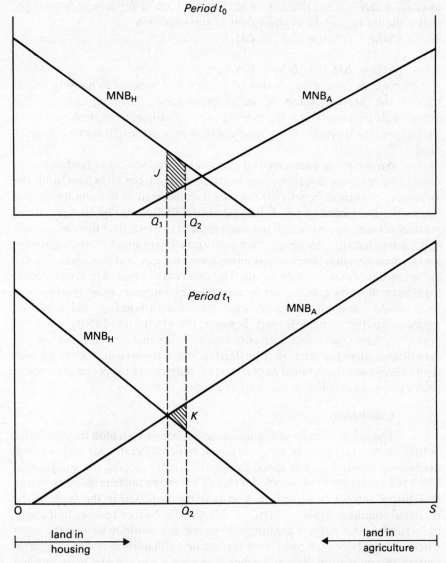

Fig. 3.9 Optimal allocation of land between competing time uses over two time periods.

housing land is expanded to Q_2 the net gain in current benefits is shown by the shaded area J[29]; the net loss in future benefits (which will be the appropriate discounted measure if the curves shown are already in discounted form) is given by the shaded area K. As drawn, J is greater than K and therefore it is worth while transferring at least Q_2 units from agriculture to housing use in the current period. If we indicate total benefits from agricultural land use as A and from housing use as H, with subscripts zero or one to indicate current or future periods; then the shaded area J may be defined as $(\Delta H_0 - \Delta A_0)$, and K defined as $(\Delta A_1 - \Delta H_1)$. The criterion for shifting land out of agriculture beyond Q_1, that J should exceed K, is equivalent to the condition

$$(\Delta H_0 - \Delta A_0) > (\Delta A_1 - \Delta H_1)$$

or

$$(\Delta H_0 + \Delta H_1) > (\Delta A_0 + \Delta A_1)$$

In other words, the present value of incremental benefits in housing should exceed the present value of incremental losses from agriculture. Not surprisingly perhaps, this is the essence of the cost-benefit analysis procedure as it is typically applied to the assessment of change in use of a discrete parcel of land.

An optimum intertemporal and inter-use allocation of land overall is defined by applying this procedure to successive units of farm land, until the increment of current benefit (J) from land transferred to housing finally just equals the increment of lost future benefits (K) from foregone agricultural production services. Although our diagram is drawn to reflect this cost-benefit approach to land use decisions, it is equally possible to structure the analysis in a way which parallels the previous models where a user cost curve is derived to guide current resource use decisions. The user cost of a unit of land transferred from agriculture in t_0 is the net benefit thereby foregone in t_1. Positive user costs would commence after Q_1 were allocated to housing, and would be measured as the vertical distance between the MNB_A and MNB_H curves of period t_1. Combining this user cost curve[30] with that showing the current agricultural benefits lost in transferring land to housing in t_0 enables determination of the optimal land allocation pattern (and appropriate transfer price) to be established in the current period.

Conclusion

The prime purpose of this discourse has been to establish the criteria for defining a natural resource use policy, and to demonstrate that such a policy necessarily requires a clear specification of current resource use and pricing. The basic condition to be satisfied is one of balancing the true social costs and benefits of current consumption, and may be expressed in the form of the marginal condition $MSB = MEC + MUC$. This further reduces to the simple statement that current consumption should not continue beyond the point where the net benefit derived from the last unit equals its user cost. A central feature of any natural resource policy is pricing – and the key to an optimal pricing policy is the correct measurement of the intertemporal social costs of resource consumption. Since user costs are in the nature of an externality, it almost inevitably demands intervention into market processes to ensure that they are 'internalised' and thus appropriately charged to the current generation of consumers. This suggests the relevance of an array of market measures from taxation and nationalisation at the national level, to price fixing, supply contol and producer groupings at the international level.

For a variety of reasons the suggested classification of natural resources will not provide a definitive set of boxes into which any particular resource can be uniquely placed. However, arguments over the exact definition and typology of natural resources provide little more than an intellectual exercise for the tidy mind, so this should not be considered of great consequence. Rather it is important to recognise that the classification highlights the four distinct variations on the economic decision problem that society faces in the optimal intertemporal use of its resource stocks, and therefore provides a basis for the introductory treatment of resource economics. Onto this structure can then be grafted the no less important discussions of imperfections and failures in resource markets with the associated problems of externalities, public goods, property rights, etc. which are central to this area. Much more could be said to shift 'depletion' from the realms of the emotive to the realistic, to examine the arguments for the social ownership of natural resources and the need for integrated resource use policies which recognise the substitution between different resources and the inevitable dilemmas of conservation. These, however, are the natural resources retained for exploitation in the next period.

Notes

1. To the purist perhaps, stocks of *all* resources are subject to natural rates of change over time, since presumably new land areas, new ores, coal seams etc., are being formed somewhere on the planet all the time. However, the rate of change is so imperceptibly slow as to be irrelevant, if we accept that geological time scales and economic planning horizons are not comparable.
2. Although, as we shall see, the full consideration of natural resource use inevitably involves production-type decisions as well.
3. Again, hard-and-fast distinctions are not always tenable. It so happens, for example, that the supply of some natural resource services represents a flow (e.g. solar energy); furthermore, social policies towards education and investment in human capital might be regarded as an intertemporal reallocation of labour services.
4. If such myopia is difficult for the reader to accept, no great difficulty is presented by adopting an alternative assumption that some specified quantity of the stock is to remain at the end of the planning horizon, with the problem being to optimally allocate the quantity earmarked for consumption within the planning horizon.
5. The level of abstraction embodied in the conventional demand and supply curves, for example, wherein a vast segment of the real world economic environment is excluded by holding 'everything else except price' constant, is huge. None the less, simple demand/supply analyses still provide a powerful framework for gaining understanding and drawing realistic inferences about market behaviour.
6. Witness the existence of active conservation (and preservationist) movements, which are indicative of group pressure to value future consumption more highly against current use; also the adoption of relatively low 'social' rates of discount in many public projects.
7. Which is similar to that used by Heady (1950), for example, in his discussion of conservation.
8. In the light of this, the decision by the OPEC countries to suddenly but belatedly raise oil prices in 1973 on the premise that 'oil in the ground is worth more than oil in the barrel', far from being just an expression of market power, could be seen as an important step in shifting towards a more rational world energy policy. From the standpoint of a global society, it was the extent and suddenness of the adjustment that were unacceptable, not the fact of it.

9. The marginal net benefit from consumption in any period is simply the difference between MSB and MEC.

10. Note the possibility of a distinction between the *physical stock* of the resource and the *useful services* that the resource provides (in this case the two are identical). From the point of view of society, it is the allocation and consumption of resource services that is the prime concern, and this significantly affects the meaning that we attach to the 'use' of natural resources.

11. The diagrams have been drawn on the purely arbitrary basis that both the MSB and the MEC curves are identical in each period. However, since future marginal costs and benefits must be appropriately discounted to their present values before comparison with the equivalent current quantities, the discounting process causes MSB_1 and MEC_1 to be 'lower' than their counterparts in t_0.

12. This is not to imply that 'leaving sufficient to satisfy future demand' is a meaningful guideline in resource policy. Clearly, however, current consumption at a level of $O\bar{Q}_0$ cannot be an optimal intertemporal allocation, for the net benefit obtained from the last unit consumed in t_0 is (zero) far less than that gained from the final consumption unit in t_1.

13. This implies a sort of ultimate 'stationary state'. Clearly to cater for growth in resource demand over time requires reserve stocks of the natural resource.

14. Many interesting and important questions revolve around these particular asumptions which open the way for a number of variations in the analysis. However, they cannot be explored in an introductory treatment such as this.

15. Where severe limitations are faced on resource availability user costs may be incurred from the outset.

16. This is similar to the familiar treatment in market price discrimination theory of summing marginal revenue curves of separated markets to yield the aggregate marginal revenue curve.

17. Although coal, oil resources and guano, for example, originate from biological sources, the passage of time has transformed them out of this category.

18. It would seem that the technological intervention of man can adjust the rate of regeneration of some biological stock resources so that r is not totally exogenous. For example, 'artificial' fish farming, afforestation – indeed the whole apparatus of modern agricultural production – seem to suggest a technological control of biological processes that is almost akin to recycling. On the other hand, it could be argued that this is merely encouraging the achieved regeneration rates to approach their fundamental biological limits, rather than changing those limits. Having dealt with the 'pure' recycling case we shall avoid these quicksands by concentrating on 'pure' biological renewal – as typified by the exploitation of marine fish stocks, for example.

19. This is a type of graphical construction which will be found most useful also for exploring the implications of parameter changes in Type I resource situations.

20. Compare this with Nautiyal (1970) who incorrectly argues that natural regeneration results in *lower* user costs. Admittedly he was dealing with forest investment and the replanting of timber after felling, and seemed to be handling the problem in the context we have treated as recycling; however, as we have seen there are some fundamental differences between recycling and natural regeneration such that the terms should be strictly associated with particular types of resources.

21. More correctly perhaps, where the rate of regeneration is dependent upon the size of the residual stock (with marine fish stocks, for example) this quadrant should reveal an 'S'-shaped curve.

22. The reason for this is most easily shown algebraically. If we let q_0 and q_1 represent the units consumed in the current and future periods respectively, then $B_0 = f_0(q_0)$ and $B_1 = f_1(q_1)$ are functions expressing the *total* net benefits derived from consumption in each period (these would be quadratic in form in the case shown in Fig. 3.7, where the respective *marginal* net benefit curves are linear). By definition,

the marginal user cost of current consumption is the loss of future benefit (ΔB_1) incurred by an additional unit of current consumption (Δq_0), or (minus) dB_1/dq_0. By the rules of implicit differentiation,

$$\frac{dB_1}{dq_0} = \frac{dB_1}{dq_1} \cdot \frac{dq_1}{dq_0}$$

1. The first component of this expression, dB_1/dq_1, is the marginal net benefit of future consumption MNB_1.

2. The second component, dq_1/dq_0, is the amount by which future resource availability alters as a result of current consumption. This has a value of $-(1 + r)$, since by definition $q_1 = (S - q_0)(1 + r)$.

   The MUC curve is thus defined as $(1 + r) \cdot MNB_1$. Since in Fig. 3.7 MNB_1 is a linear function of the form $a - bq_1$, then MUC can be written as $a(1+r) - b(1+r)q_1$. However, this expresses MUC as a function of q_1 whereas it needs to be expressed in terms of q_0 if it is to be included as part of the social costs of current consumption. Substituting for q_1 yields
   $$MUC = a(1+r) - b(1+r)\{(S - q_0)(1+r)\} = a(1+r) - b(1+r)^2 S + b(1+r)^2 q_0$$
   This is a linear function of the form $K + b(1+r)^2 q_0$; its slope, $b(1+r)^2$, is thus $(1+r)^2$ times greater than the slope of MNB_1^2.

23. The validity of this optimality criterion can be shown alternatively by referring back to Fig. 3.1. In the case of a resource stock which grows from one period to the next the interperiod consumption possibility line would have a slope of $(1+r)$, and optimality is defined when this is equated with the slope of the indifference curve (being the marginal rate of substitution in consumption between t_0 and t_1). This marginal rate of substitution, dC_1/dC_0, is by definition the inverse ratio of the relevant marginal utilities, or in this case MNB_0/MNB_1. Hence an optimal intertemporal allocation of the stock is determined when
   $$\frac{MNB_0}{MNB_1} = (1+r).$$

24. This is what, in essence, much of the land use methodology in agricultural economics is concerned with—for example the determination of an optimal land use pattern (e.g. cropping programme) for a defined production period.

25. To the extent that these resources enter the market system.

26. For a particular area of land, its location, climate and topography—inasmuch as they are unchangeable—are also part of its space characteristics.

27. Although some urban land could possibly be returned to agricultural uses, the capital and reclamation costs are so high as to make this reversal effectively infeasible.

28. Note we are here abstracting from the 'extraction' costs of these services by deducting them out from MSB in order to arrive at a (net) measure of benefits that is comparable between uses.

29. Since the area under a marginal benefit curve measures total benefit, the shaded region J represents the difference between the added benefit from additional housing and the reduced benefit from lost agricultural production.

30. Note that as defined this is the 'user cost of housing land'.

References

Allen, S.W. (1955) *Conserving Natural Resources: Principles and Practice in a Democracy*, McGraw-Hill, New York.

Bain, J.S. (1973) *Environmental Decay*, Little, Brown, Boston.

Barclay, P.W. and **Seckler, D.W.** (1972) *Economic Growth and Environmental Decay*, Harcourt Brace Jovanovich, New York.

Barnett, H.J. and **Morse, C.** (1963) *Scarcity and Growth: The Economics of Natural Resource Availability*, Johns Hopkins Press, Baltimore.

Ciriacy-Wantrup, S.V. (1956) *Resource Conservation: Economics and Policy*, University of California Press, Berkely.

Clapham, J.H. (1922), 'Of empty economic boxes', *Economic Journal 35*, pp. 305–14.

Dales, J.H. (1968) *Pollution, Property and Prices*, University of Toronto Press.

Dasgupta, P. and **Heal, G.M.** (1974) 'The optimal depletion of exhaustible resources' in *Symposium on the Economics of Exhaustible Resources Review of Economic Studies* pp. 3–28.

Edel, M.A. (1973) *Economies and the Environment*, Prentice-Hall, New Jersey.

Ehrlich, P.R., Ehrlich A.H. and **Holdren J.P.** (1973). *Human Ecology: Problems and Policy*, Wiley, London.

Ely, R.T. and **Wehrwein, G.S.** (1940) *Land Economics*, University of Wisconsin Press, Madison.

Fisher, I. (1930) *The Theory of Interest*, Macmillan, New York.

Freeman, A.M., Haveman, R.H. and **Kneese, A.V.** (1973) *Economics of Environmental Policy*, Wiley, London.

Gordon, R.L. (1957) 'A re-interpretation of the pure theory of exhaustion', *Journal of Political Economy 75*, 274–85.

Gray, L.C. (1912) 'The economic possibilities of conservation', *Quarterly Journal of Economics 27*, 497–519.

Hardin, G. (1968) 'The tragedy of the commons', *Science 162*, pp. 1243–8.

Heady, E.O. (1950) 'Some fundamentals of conservation–economics and policies', *Journal of Farm Economics 32*, 82–95.

Herfindahl, D.C. and **Kneese, A.V.** (1974) *Economic Theory of Natural Resources*, Merrill, Columbia, Ohio.

Hotelling, H. (1931) 'The economics of exhaustible resources', *Journal of Political Economy 39*, pp. 137–75.

Jarrett, H. (Ed.) (1966) *Environmental Quality in a Growing Economy*, Johns Hopkins Press, Baltimore.

Keynes, J.M. (1936) *The General Theory of Employment, Interest and Money*, Macmillan, London. Appendix Ch. 6.

Laidler, D.A. (1974) *Introduction to Microeconomics*, Philip Allan, Oxford, ch. 8.

Meadows, D.L. *et al* (1972) *The Limits to Growth*, Earth Island, London and Universe Books, New York.

Nautiyal, J.C. (1970) 'User cost in renewable natural resources', *Canadian Journal of Agricultural Economics 18*, pp. 95–103.

Nicholson, W. (1972) *Microeconomic Theory: Basic Principles and Extensions*, Dryden Press, Hinsdale, Illinois.

O'Riordan, T. (1976) *Environmentalism*, Pion, London.

Scott, A.D. (1953) 'Notes on user cost', *Economic Journal 63*, pp. 368–84.

Scott, A.D. (1955) *Natural Resources: The Economics of Conservation*, University of Toronto Press, Toronto.

Solow, R.M. (1974) 'Intergenerational equity and exhaustible resources' in Symposium on the Economics of Exhaustible Resources, *Review of Economic Studies* pp. 29–46.

Tomkins, J. and **Butlin, J.A.** (1975) 'A theoretical and empirical approach to fisheries economics', *Journal of Agricultural Economics 26* pp. 105–25.

Part Two

Time, like an ever-rolling stream,
Bears all its sons away.   *Isaac Watts*
Intertemporal and intergenerational problems

Introduction _____

Encapsulated in environmental and natural resource management is the problem of time. It permeates each resource utilisation problem with which man is faced. 'Neoclassical' capital theory incorporates the time dimension in a quite adequate way, from the point of view of the current generation, evaluating intertemporal resource allocations using the equivalent present value of alternative allocations (at current real rates of interest) as the appropriate choice set and from which the allocation with the highest net present value is the socially optimal intertemporal allocation viewed from the present. But can we trust myopic, efficiency-based, self-interest maximising criteria to allocate resource use between generations, particularly if the allocation is from a fixed stock of exhaustible resources? Is the current generation convinced that decisions about exhaustible resource extraction made earlier in the twentieth century are appropriate from the viewpoint of the present generation. (Oil is the example that springs most immediately to mind.) This leads us not only into a discussion of efficiency versus equity between generations, but also into a discussion as to whether the utilitarian, present-value-maximising criterion is only myopically efficient.[1] Most of the section is concerned primarily with the 'fairness' of alternative rates of extraction resulting from alternative choice criteria for alternative rates of extraction.

Professor Heal's inaugural lecture gives an excellent elementary but thoroughgoing introduction to the issues involved. (It will be useful to refer back, on occasion, to McInerney's paper, in which the allocations are determined largely on a utilitarian decision rule.) An alternative, equity-based decision rule is discussed, and the implicit rates of extraction compared. Professor Solow (1974), evaluating the Rawlsian rule more thoroughly under a restrictive set of conditions, determined that optimal extraction, with a Rawls-type maximum approach, involved equal sharing of the finite stock of an exhaustible resource over a finite number of generations. Heal is more concerned here with setting out the fundamentals of the two approaches, and intuitively evaluating which approach appears to guide those who make decisions about exhaustible resource extraction.

Professor Page is equally concerned about intergenerational equity problems relating to resources use, but broadens his interest to embrace secondary as well as virgin materials. He argues comprehensively both here and elsewhere (Page, 1977, intro.) that resource use decisions ought to be based on all available resources, and not simply virgin sources. For a variety of reasons, the market system may be biased towards extraction from virgin sources, and neglect the resource base that the accumulated stock of waste materials represents. The possibility of expanding secondary materials recovery whilst maintaining the resource base intact (Page, 1977, pp.12–14) is a policy option that Page has developed, and which appears to have more operational relevance than other, equity-based intertemporal decision rules to determine the time-path of exhaustible resource depletion.

Paul Grout's paper has quite a different pedigree from the other two in this section, and serves quite a different purpose.[2] It turns the reader's attention resolutely back to the body of microeconomic welfare theory, and reminds him that it is from this theoretical basis that much of the more applied concerns expressed elsewhere in the book derive their underpinnings. In the light of the excellent developments of some of the fundamental aspects of intergenerational equity problems in the papers by Heal and Page, Grout's development of the criteria used enables the interested reader to follow the subject further without requiring too high a level of specialist economic knowledge.

Notes

1. For a full discussion of this see R.H. Strotz: Myopia and inconsistency in dynamic utility maximisation, *Review of Economic Studies 23* (1955/6), pp.165–80.
2. Those who are not specialist economists may need to seek assistance to derive the full benefit from this paper.

References

Solow, R.M. (1974) Intergenerational equity and exhaustible resources in Symposium on the Economics of Exhaustible Resources, in *Review of Economic Studies*, pp.29-46.
Page, Talbot (1977) *Conservation and Economic Efficiency*, Johns Hopkins Press/Resources for the Future, Baltimore.

Chapter 4
Economics and resources*   *G.M. Heal*

Although the title of this chapter, 'Economics and resources', is a seemingly clear phrase, it may in fact mean different things to different sectors of the audience. Non-economists will no doubt expect a disquisition on what economists have to say about the use and depletion of the earth's resources of oil, coal, mineral ores, etc: to the general public, the word resources immediately conjures up issues of this type. To the economists, however, it has a slightly broader import: the subject matter of economics was defined in the 1930s as 'The study of the allocation of scarce resources between competing uses', and this is a statement that is still difficult to better as a simple definition. But when in this definition we refer to the allocation of scarce resources, we have in mind not just the earth's exhaustible resources, but rather the whole of society's economic resources – i.e. all possible inputs into the productive process. This of course includes exhaustible resources, but also contains available inputs of human skill and capital equipment, and in general when economists talk about resources and resource allocation the word is being used in this broad sense to include both earth's resources and the human and capital resources available to a society.

Taking resources in this broad sense, economists have traditionally been much concerned with defining what is meant by the efficient, or indeed optimal, use of resources, and with questioning whether the unhampered play of market force will achieve such a state; if not, why not, and what forms of government intervention are in consequence most necessary and appropriate.

Though much intellectual energy has been devoted to the issue of how society uses its resources, in the extended sense of the word, and how those use patterns can be altered, it is fair to say that those involved in this work, at least this century, have always had at the back of their minds an interpretation of the word resources that more or less restricts it to capital and labour: certainly most of the concrete examples they have chosen to study have involved capital, or labour, or both. Of course, this is not surprising when one considers the conditions under which this theory evolved: during the 1930s the dominant economic problem was clearly the apalling misuse of society's labour resources, and during and after the war it has generally been seen as a shortage of, or the existence of an outdated stock of, capital equipment. To find examples of resource-allocation theorists explicitly concerned with the use of the earth's exhaustible resources as exemplars of their problems, one has, bar one notable exception Harold Hotelling (1931), to go back to the nineteenth

---

*Professorial Inaugural Lecture, University of Sussex, 1974.

century. Again, this is not surprising: in the eighteenth and nineteenth centuries, the ownership and productivity of land were clearly of great importance in determining the distribution of income, and the timing and location of the industrial revolution in the UK were greatly influenced by the exhaustion of traditional resource supplies and the availability of others. So to nineteenth-century man it would have seemed unthinkable that one could explain the dynamics of an economy, or analyse the process of production and exchange, without giving special attention to the role of natural resources. Yet, amazing though this would seem to nineteenth-century man, this is just the procedure adopted by most twentieth-century economists – presumably reflecting the fact that for the first two-thirds of this century, exhaustible resource constraints have not been important for industrialised countries. These either possessed their own supplies of resources, which they felt to be adequate, or felt that they could be confident of importing them in unlimited amounts from developing countries, initially in many cases because they controlled these as parts of the colonial system, and subsequently because, though independent, the supplying countries remained politically quiescent, with foreign exchange needs so great that they could be relied on to export unlimited amounts of their principal (and often only) exports.

One of the points that emerges from this brief review is that though economists have traditionally been very concerned with the allocation of a society's resources, the particular category of resource that they have chosen to focus attention on has varied in response to the economic and social conditions of the times, from natural resources in the nineteenth century, to labour in the 1930s and capital subsequently. Naturally the wheel has now turned full circle: increasing intellectual effort is being focused on issues raised by the existence of exhaustible resources.[1] This chapter reviews some of these.

In so doing, there is clearly an immense problem of selection. This has been resolved by focusing on the underlying intellectual problems in this field.

When one is discussing the use of exhaustible resources, there is one basic issue that dominates all others, at least when one takes the long, intellectual view: it is, quite simply, *how fast should we use them up*? For something which is important, which is destroyed in use, and which is available only in limited amounts, we *have* to take a position on this question. Clearly some groups, notably conservationists, feel that we are using resources 'too fast' – and for the time being this phrase just has to be in inverted commas. On the other hand, it is more or less explicit in much political and press discussion of these matters that the sooner we can use the resources the better. For them, there is no 'too fast', just 'the faster the better'. Who is right here?

With a view to resolving issues of this type, one of the matters that economists have been concerned with is to see whether one can produce an intellectually defensible yet practically useful definition of an 'optimal depletion rate', a rate of consumption which is better than all others, and is neither too fast nor too slow.[2] You will appreciate immediately how ambitious a task this is, and clearly one would not necessarily be confident of success: but simply setting out the issues involved in defining such a rate can make a useful contribution to clarifying the debate in this area – and certainly one can quickly say that some suggested policies could never be optimal – amongst the 'the faster the better' policy.

What are the factors relevant in choosing between depletion rates, and selecting an optimal depletion rate? Most fundamental are considerations of equity – not of equity between those currently alive, but of equity between present and future, between current and subsequent generations. Clearly, the faster we consume exhaustible resources, the less is left for our children and our children's children, and so, other things being equal, the lower will be the living standard they can achieve. Thus, over the long run, we have to strike a balance between our own wishes and those of our successors. This clearly involves some analysis of what might seem equitable in this field. Note that there is an important difference between this problem of intergenerational economic equity, and the problem of achieving economic equity between those currently alive, who are members of the current generation. This is that the poor of the present generation are alive and can press their case: obviously the rich may choose not to act in response to this pressure, but they cannot be entirely unaware of the problem. But future generations, whose well-being is at stake in the intergenerational problem, have no way of presenting their case: and if the rich of today can largely neglect the present poor, who are here to press their case, then the future will need very articulate spokesmen if they are to obtain a fair hearing.

Returning to the issue of equity, there are two frameworks within which economists have worked when approaching this, the Utilitarian and the Rawlsian – the latter named after John Rawls, the present Professor of Philosophy at Harvard, whose work *A Theory of Justice* is probably known to many of you. The Utilitarian approach would suggest that a depletion rate ought to be chosen so as to maximise the sum of the benefits from resource use accruing to all generations, present and future. Though sounding rather abstract, this is the kind of approach that can be made operational within the conventional framework of cost-benefit analysis, and indeed has been made operational in this way on many occasions. One of the issues that has to be faced in thus making it operational, is how to weight the various benefits accruing at different dates when adding them up. For example, if we have benefits of 100 units in each of the next thirty years, do we declare their sum to be 30 x 100? Or do we adopt a more complex approach, giving different weights to benefits in different years? If we were to follow the kind of approach which individuals seem implicitly to follow in making their own economic decisions, then we would give benefits occurring in the future less weight than those occurring at present, the differential in the weights increasing with the degree of futurity. This is a practice economists refer to as *discounting* future benefits, and it is an approach that most individuals implicitly adopt: the evidence for such a statement is, roughly speaking, that if I ask you which of the following two alternatives you would prefer, £100 today or a sum chosen to have exactly equivalent purchasing power a year from now (it would be the same sum in the absence of inflation), you would almost certainly prefer the present sum: in making such a choice, you are in effect saying that the further away from the present a sum is, the less valuable it is – you are discounting future benefits. Now the following question has been discussed extensively: if every individual in a society discounts future benefits in the way I have just indicated, then should a government making decisions on resource depletion rates, and acting for the totality of those individuals, adopt the same procedure? This is an issue on which many different positions have been adopted: Frank Ramsey, a Cambridge economist and philosopher writing in the 1920s, declared it to be his view that 'discounting of future benefits is ethically indefensible and arises purely from a weakness of the imagination' (Ramsey, 1928). Pigou, a Cambridge economist of the same period, shared his approach: he argued that individual decision-making processes with regard to the future were myopic, and suffered from a 'faulty telescopic faculty'. The government's role was to take a stance which counteracted this short-sightedness of its citizens, thus making the whole better than the sum of its parts. Nowadays, however, not everyone would agree with this prescription that the government should give benefits in all periods equal weight; and this for several reasons. One, that has been applied particularly in the context of developing countries, is that income levels may be expected to rise over time: the future people who receive future benefits will be richer than the present people who receive current benefits, and therefore less deserving of our sympathy and consideration, and we should discount the benefits accruing to them at a rate depending on the anticipated income differential between them and us. An example of this point would be the observation that in retrospect it would seem tragic if those who were relatively poor in the earlier part of this century had been asked to take further cuts in their living standards simply to benefit those who would be alive in the latter part of the century, and who anyway have so many advantages over them. If we give great weight to future benefits, and therefore undergo some privations in order to ensure these, it is possible that their rich recipients will feel towards us as we would feel towards our poorer predecessors if they had undergone unnecessary deprivation in our interests. Of course, it is implicit in this argument that if income levels are expected to fall, rather than rise, over time, then we should give more, rather than less, weight to the future – we should do the reverse of discounting. And of course many of you will have heard of forecasts, by the Club of Rome and others, suggesting that living standards will indeed fall as pressures on resources rise.

One reason has now been mentioned as to why not everyone would agree that present and future benefits should be given equal weight: this has to do with differences between present and future incomes. A second and more technical reason has to do with uncertainty. Consider the hypothetical experiment in which I ask you whether you would prefer £100 today or a sum of equivalent purchasing power in a year's time. Suppose that you choose the £100 now, and we then ask you to justify your choice: after a certain amount of confusion, you might perhaps refer to the old adage that a bird in the hand is worth two in the bush. If one interprets the hand as the present and the bush as the future, and there is substantial uncertainty about what the future will bring by way of technological changes, new resource discoveries, war and devastation, etc., then it may indeed be the case that a bird in the hand is worth, if not two in the bush, then at least substantially more than one. This point has been made more precise, and indeed operational, by reference to models of choice under uncertainty developed by statistical decision theorists:[3] however, it is appropriate here simply to note that uncertainty about the future may provide a valid reason for discounting future benefits.

Thus far the discount rate, or equivalently the weights, to be used in summing the benefits occurring at different points of time under alternative resource-depletion policies have been discussed. Obviously, rapid depletion will lead to a concentration on present benefits at the expense of future ones, and will be favoured by a high discount rate, and vice versa. This utilitarian approach to defining an optimal depletion rate would then have us choose that policy giving the highest sum of appropriately-discounted benefits. The role of discount rates is worth emphasising, because the Utilitarian approach is the one usually chosen by governments, and whenever they operate it they have, explicitly or implicitly, to choose a discount rate. The UK government does this quite explicitly, and chooses a rate of about 10 per cent sometimes a little higher. This implies that 10 per cent is deducted from the value of a benefit for each year of futurity, and the destructive power of such a rate is remarkable: if one discounts at 10 per cent for each year of futurity, £100 in 1984 is worth only £37 today, by 1994 it is worth only £13.50, and by the end of the century it is hardly worth having, at £7.42. Do you really believe that £100 to our children is only worth £7.42 as far as we are concerned?

The particularly close focus on discount rates has been chosen because, within the economic decision-making framework usually used, they are a key variable, the pivot about which we strike a balance between present and future: yet because they are very technical, choices in this field usually go unchallenged. More understanding and discussion can only be helpful: how many of you, for example, would have voted in the last election for a party promising to give effectively zero weight to considerations relating to resource stocks and the environment at the end of this century? And how many politicians, when they approve a departmental recommendation for a 10 per cent discount rate, really understand the implications of their choice?

We have now seen some aspects of the Utilitarian approach to defining an optimal depletion rate: it is that rate which gives the highest sum of the appropriately-discounted benefits occurring at various points of time. Obviously, different discount rates imply different depletion rates: given certain assumptions about the technology for extracting and consuming resources, it is possible to sketch the general characteristics of an optimal depletion policy for any discount rate, and to say how this policy varies with the discount rate.[4] At that point, the economist's role is complete: the choice of one discount rate rather than another is not an economic choice, but a political and moral one. Economists should perhaps point out the consequences of different choices, but it is up to society as a whole, or its representatives, to make the choices.

What has been said of the Utilitarian framework is probably now enough to indicate the essence of the approach, and in particular to indicate the key issues and the forms in which they arise: the next step is clearly to do likewise for the alternative, Rawlsian approach. Rawls, in his unusually ambitious book *A Theory of Justice* (1972), enunciates, and attempts to justify and spell out the implications of, what he regards as the two principles of justice. These are that:

> Each person is to have an equal right to the most extensive total system of equal basic liberties compatible with a similar system of liberty for all.

This is a widely-acceptable nostrum, and certainly not one that is at issue here. More relevant is the second principle, which requires that:

> Social and economic inequalities are to be arranged so that they are both
> (*a*) to the greatest benefit of the least advantaged, consistent with the just savings principle, and
> (*b*) attached to offices and positions open to all under conditions of fair equality of opportunity.

Rawls's justification for these principles is interesting. In essence, he argues that people's conceptions of a just society are heavily conditioned by their positions in society, so that to obtain information about what one might describe as a more objective concept of justice, one has to conduct the hypothetical experiment of asking what kind of society people would regard as most desirable if they were completely uninformed about the position they would occupy in that society. In other words, your true views on what constitutes social justice are not those that you subscribe to, given that you occupy a rather privileged position in society, nor are they those that you would subscribe to if you were in an under-privileged position: they are those that you would support if your true position in the society was unknown and yet to be determined, and was as a result hidden from you by what Rawls calls a 'veil of ignorance'. He then argues that, if making their choices from such a detached position, people would always insist that, to quote his second principle, 'Social and economic inequalities are to the greatest benefits of the least advantaged . . . ' Probably the outline of the reasoning leading to this conclusion is clear to you: it depends upon what economists call 'risk-averse Behaviour' or, to put it more classically but no less cynically, upon the (almost) Shakespearean maxim that '[Social] conscience doth make cowards of us all'. But rather than go into the details of the argument, let us look more carefully at the principle: 'Economic inequalities are to be arranged so that they are to the greatest benefit of the least advantaged' – a statement in which many of you will see obvious analogies to the maxim choice rule of games theory. Indeed Rawls's principle is often referred to as the maxim principle of justice, because it identifies justice with institutions and policies designed to maximise the well-being of those with minimum well-being, and implies that inequalities are justified only in so far as they contribute to the well-being of the less fortunate. This is an interesting criterion, and clearly one that can be given operational meaning within the intergenerational context with which we are concerned. But before doing that, it should be noted that Rawls himself does not see this principle as applicable in an intergenerational context: he does in fact address himself directly to the issue of intergenerational justice and equity, and proposes what he calls the 'just savings principle' as a principle which should govern the treatment that each generation should mete out to its successors. This principle is very imprecisely defined, and unfortunately most interpretations of it suffer from the substantial disadvantage that they clash with other more obvious and more intuitively acceptable criteria with which a savings principle should comply, so that most economists have chosen not to pursue the implications of this principle, but instead to use the original maxim principle. This suffers from no such shortcomings, and its justification seems equally acceptable in either the intertemporal or the static contexts.

Pursuing this approach, then, an optimal resource-depletion rate is one so chosen as to maximise the benefits form resource-use that will accrue to that generation who will receive the smallest such benefits: in other words, which maximises the benefits accruing to that intertemporal group which will receive minimum benefits. Now, fairly obviously, the identity of the group which will receive minimum benefits from a depletion policy will change as that policy changes: under very rapid depletion policies, it will be a far-distant group, but under a very strict conservation ethic it might be the present generation. It follows from this that in identifying an optimum, in this Rawlsian sense, one has to follow a two-stage procedure: firstly, for each possible depletion policy, identify the least-advantaged group, and, secondly, then select that policy whose least-advantaged group is best treated. In case this should sound unpleasantly complex, and indeed so complex as to violate by earlier requirement that our approaches should yield operational criteria, let me reassure you on this point. It can be shown that if we carry out this two-stage search procedure, then what we will eventually identify is in fact beautifully simple: we shall in the end select that depletion rate which gives the highest *sustainable* level of consumption of those outputs for which resources constitute essential inputs. Paraphrasing this, we shall select the highest-possible *steady* consumption rate in those areas dependent on resource-use. Although the precise manner in which this rabbit emerges from the hat may not be clear – reasonably so, as the detailed argument is very complex[5] – it is perhaps not too difficult to grasp the point intuitively: the Rawlsian criterion manifests a strong concern for intergenerational equity, and its implication is that all generations should be treated equally – for this is of course what is involved in choosing a steady, sustainable consumption path.

Having presented the two basic approaches, Utilitarian and Rawlsian, it will be beneficial now to step back from them a little and see how well they accord with other ideas that have been expressed in this area, in particular those expressed by people who have taken an unambiguous stand on the optimality or otherwise of our present resource-usage rates. Prominent amongst these are those who have contributed to the Club of Rome's grandiloquently-titled 'Studies of Man's predicament', (Meadows, 1972) and to the 'Blueprint for Survival' (*The Ecologist*, 1972). If they regard present depletion rates as 'too high', perhaps they have in mind an optimum which we err in exceeding – and perhaps some at least implicit definition of this optimum can be found in their writing? Sadly, this line of argument seems to be false: it seems that those who make the most strident statements about depletion policies being 'too fast' or 'too slow' are quite remarkably successful at hiding the criteria according to which they make such strong judgements. Only in one place can one find an explicit statement of such criteria, in a paper written by the authors of the Club of Rome report 'Limits of Growth' after the publication of their report. They (Randers and Meadows, 1973) argue that we should accept a

> long-term objective function that maximises the benefits for those living today, subject of course to the constraint that it should in no way decrease the economic and social options of those who will inherit this globe.

It is not altogether easy to give precise interpretation to this idea, but it would seem fair to interpret it as requiring that we maximise the present generation's consumption from resources, subject to the limitation that we do not in so doing force succeeding generations to lower levels of resource-dependent consumption. But the chances of their reaching higher levels of consumption than us, if we do the best we can for ourselves subject to this limitation, are very low: hence once again this amounts to picking the highest *sustainable* resource-dependent consumption level. So it seems fair to say that at least one of the two frameworks economists have used in analysing what is meant by 'optimal' resource depletion, is in general terms in keeping with the opinions expressed by non-economists interested in the area.

After talking for some time at the level of abstractions and principles, it is now appropriate to turn to some more specific issues, and in particular to address questions such as:

> 'Can the kinds of general principles enunciated so far be made to yield moderately precise information about the rates at which we should deplete resource stocks?' and 'To what extent is the kind of economic system within which we operate geared to achieving depletion rates that are not too dissimilar from those that our principles would recommend?'

First consider the former of these issues – the extent to which one can use these abstractions to derive numerical estimates of optimal depletion rates. Only the Utilitarian approach is sufficiently well-developed to do this, and within this context one can identify two variables that play a key role in determining the value of the optimal depletion rate. It follows rather clearly from what I have already said that one is the discount rate: the higher this is, the less is the weight given to future benefits and the greater the depletion rate that should be chosen. The second important variable is perhaps a less obvious one, and one that I have not yet mentioned: it is a variable describing the rate at which consumers become satiated by consumption of the goods that the resource is used to produce.[6] Being satiated means that they have sufficient of these goods and that more are of little or no value to them. Clearly we do rapidly become satiated by the consumption of any particular good – there is, for example, a limit to the number of cars or televisions that we can usefully own, or to the amount of food we can eat with pleasure and advantage. While it is clear that people can reach satiation levels for the consumption of any particular good, it is less clear that they become satiated with the consumption of goods in general. But that is an aside: for present purposes it suffices to note that the more rapidly people reach satiation levels of consumption for goods that are produced from resources, the lower will be the optimal depletion rate. This is almost a self-evident truth: the faster satiation levels are reached, the less are the justification and incentive for extensive resource consumption in any period, and so the lower should be the rate of use.

There are two parameters – the discount rate and the rate at which satiation is reached – which one would, at a purely intuitive level, expect to determine one's choice of an optimal depletion rate, and indeed this point can be made more formally and precisely by a rigorous analysis which would be out of place here. This means that if we can evaluate these two parameters, we can go a long way towards determining the correct depletion rate. The discount rate is not something that we measure: it is something that we choose, and its choice involves an ethical judgement about the balance that should be struck between present and future. The rate at which satiation is approached, however, is a measurable parameter, and there has been some discussion of what might be appropriate estimates for it. Armed with these estimates, one can make statements such as the following:

> If the discount rate is 10 per cent, the annual consumption of a
> resource should be in the region of 10–25 per cent of the stock.
> If the discount rate is 1 per cent, the annual consumption should
> be in the region of 1–3 per cent of the stock.

These are, admittedly, not very precise figures, and it is in the nature of the data and concepts we are dealing with that they cannot be made much more precise. But in spite of their margins of error, they give us an idea of the orders of magnitude involved. For example, imprecise though they are, they enable one to say that if we adopt a Utilitarian approach and choose any discount rate less than about 4 per cent, the rates at which we are planning to use North Sea oil are clearly too high. On the other hand, we are probably using coal too slowly. These calculations are of course complicated by the existence of considerable uncertainty about the size of oil reserves and of future finds, but in producing the numbers mentioned, some attempt has been made to take that into account.

These estimates have been mentioned not to dwell at any length on the answers to the calculations, but to show that we *can* arrive at answers, and to show the orders of magnitude involved. The second of the two questions posed was

> To what extent is the kind of economic system within which we operate geared to achieving depletion rates that are not too dissimilar from those that our principles would recommend?

This is another particularly difficult question, and is a special case of the more general question 'To what extent can one be happy with the degree of future-orientation achieved by an economic system like our own?' Within such a system, there are many forces that combine to determine the actual depletion rate, some tending to lower it and some to raise it. Prominent amongst the forces tending to reduce depletion is the existence of monopoly control of resource supplies: a monopoly resource supplier who hopes to stay in the business for a long time has every interest in making his supply last as long as possible, and this tendency is emphasised by the fact that cutting back supply raises the resource price, and hence the profitability of their activities. All of these points are exemplified in the behaviour of the OPEC countries: by exploiting their market power they have struck a resounding blow for conservation, and for the welfare of our successors.

But oil is exceptional in the degree to which its supply can be monopolised: many other resources are supplied by a range of competing producers, so that there is no monopoly power to act as a brake on consumption rates. This is unfortunate because almost all of the other forces that enter into the determination of actual depletion rates act in opposite direction to monopoly power, tending to emphasise the advantages of rapid depletion. Foremost amongst these other factors is the fact that the future is inherently uncertain, so that any decision to postpone the sale of a resource involves the owner in bearing a risk – the risk that the price in the future will be lower than the present price, because of changes in demand or the development of substitutes, or perhaps the risk of expropriation and thus loss of the resource revenue altogether. Such considerations can provide a powerful incentive for rapid depletion,[7] and except in cases of monopoly, there is no material counter to these forces. So in general one would expect these considerations stemming from uncertainty, which Keynes once referred to as 'the dark forces of time and ignorance', to result in unreasonably high rates of depletion. One should be a little more explicit about these forces, and this can best be achieved by discussing an example which shows how dependent the market outcome may be on highly subjective factors.

Consider the market for an exhaustible resource whose stocks are being depleted: as the stocks decline, it will become apparent to dealers that it will not always be possible to meet current levels of demand, and rumours of future shortage will cause the price to rise. One possible consequence of this is that traders expect the price rise to continue, and in consequence they hold stocks off the market to take advantage of the anticipated high future prices: at the same time, the expectation of enduring higher prices will cause users to economise on their demands for the good and to start searching for substitutes.

Hence future supply is increased, because traders hold goods off the market to realise high future prices, and future demand is reduced, because users begin to economise. These two movements will tend to prevent the anticipated surplus demand over supply which caused the original price rise: the problem in this particular case has therefore been self-correcting, with an anticipated shortage setting in motion forces which act to correct that shortage. But this is obviously an ideal: the way in which we would hope that markets would react. Other less satisfactory outcomes are equally possible: for example, sellers might not expect the initial price increase to continue, so that an initial price increase consequent upon rumours of future shortage could bring a wave of selling designed to take advantage of what some sellers regard as temporarily high prices. Such a reaction would be perverse, in that by raising present usage rates, it would lower future availability and simply aggravate the initial problem.

Which of these two models of market reaction better captures real world behaviour depends on how traders behave from their expectations, their best guesses, of future prices. And in assessing this, one is confronted with problems of group psychology, with the 'dark forces of time and ignorance', and with the casino-like properties of a speculative market on which Keynes and others have so often commented.

Enough ground has now been covered for it to be sensible to pause and review the journey. Initially, the varying fortunes of the economist's interest in natural resources that was intense in the nineteenth century, declined during much of this century, but has recently grown strong again. The reasons for these variations were easy to see, lying in the changing nature of the problems that society was felt to face. But the central issue in this field is clearly not a history of our attitudes to the problem, but a discussion of what might be meant by statements such as 'resources are being depleted too fast' or 'resources are being depleted too slowly', and hence also a discussion of the issues involved in defining an optimal depletion rate, a rate which by definition suffers neither from being too fast nor too slow. At least one of the approaches adopted can yield numerical answers to these questions, albeit tentative ones. Defining an optimal depletion rate naturally invites some comments about the extent to which market forces would tend to equate actual to optimal depletion rates, and in this context the role of monopoly in restraining consumption, and the role of the 'dark forces of time and ignorance' in advancing it, are of particular importance. Each of these points is on its own a substantial issue. There are some very deep and difficult problems involved in deciding on the correct rate of resource depletion, and in deciding whether our own system is likely to achieve this.

The dangers of complacency and of alarmism, though seemingly distinct, are in fact closely related, and relatively few writers in this area have avoided them completely. The attractions of being a prophet, be it of economic doom or economic redemption, are clearly great: but the social value of prophecy, as opposed to careful and dispassionate analysis of the logic and facts of the situation, has never been notable. So it is a little unfortunate that economists' dispassionate analyses, of the type discussed, have received far less publicity than some of the more apocalyptic prophecies made in this area by non-economists.

Notes
1. This point is illustrated by the collection of papers in Pearce (ed.) (1975) and *Review of Economic Studies* (1974).
2. The original works are Beckman (1974), Dasgupta and Heal (1974) and Stiglitz (1974).
3. The argument is set out in detail in Dasgupta and Heal (1974).
4. The papers in *Review of Economic Studies* (1974) conduct such an analysis.
5. It is given in detail, for a simple case, by Solow (1974).
6. This is what economists refer to as the elasticity of the marginal valuation of consumption. In the long run, this parameter and the discount rate determine the depletion rate completely, at least under reasonable assumptions. The argument for this proposition is contained in Dasgupta and Heal (1974).
7. Their role is discussed in more detail in Dasgupta and Heal (1974).

References

Arrow, K.J. (1973) *'Rawls's Principle of Just Saving'*. Institute for Mathematical Studies in the Social Sciences, Stanford University, Technical Report No. 106, September.
Beckman, J.J. (1974) 'A Note on Growth with Exhaustible Resources' in Symposium on the Economics of Exhaustible Resources. *Review of Economic Studies*, pp. 121–2
Daley, H.E. (1973) *Towards a Steady-State Economy*, Freeman, San Francisco.
Dasgupta, P.S. and **Heal, G.M.** (1974) 'The optimal depletion of exhaustible resources' in Symposium on the Economics of Exhaustible Resources. *Review of Economic Studies*, pp. 3–28.
The Ecologist (1972) 'A blueprint for survival', Penguin Books.
Heal, G.M. (1975) 'Economic aspects of natural resource depletion' in Pearce (ed.) (1975).
Hotelling, H. (1931) 'The economics of exhaustible resources' *Journal of Politcal Economy* **39**, pp. 137–75.
Meadows, D.L. *et al.* (1972) *The Limits to Growth*, Earth Island, London and Universe Books, New York.
Pearce, D.W. (ed.) (1975) *'Economic Aspects of Resource Depletion'*, Macmillan, London.
Ramsey, F.P. (1928) 'A mathematical theory of saving', *Economic Journal* **38**. December.
Randers, J. and **Meadows, D.** (1973) 'The carrying capacity of our global environment: a look at the ethical alternatives' in Daley (1973).
Rawls, J. (1972) *A Theory of Justice*, Oxford University Press.
Symposium on the Economics of Exhaustible Resources, *Review of Economic Studies* Jan. (1974).
Solow, R.M. (1974) 'Exhaustible resources and intergenerational equity' in Symposium on the Economics of Exhaustible Resources, *Review of Economic Studies*, pp. 29–46.
Stiglitz, J.E. (1974) 'Growth with exhaustible resources' in Symposium on the Economics of Exhaustible Resources, *Review of Economic Studies*, pp. 123–38.

Chapter 5
Economics of a throwaway society; the one-way economy *T. Page*

Upon our own generation lies the responsibility for passing on to the
next generation the prospects of continued well-being.
Paley Commission Report

Introduction

If you have ever visited a large mine or a large city dump, you may have
formed the impression that we are processing materials as though there were
no tomorrow. And then you may have been struck by the idea that we are
processing materials in such a hurry that there will be no tomorrow.

It is easy to agree that a high standard of living requires enormous
quantities of materials and energy flowing through the economy. Some,
however, go further by suggesting that the standard of living is in direct
proportion to the flows of materials and energy through the economy. If this
were true, a more conservative policy on materials and energy would lead to a
sacrifice in the standard of living. This view is too simple, too strict, and too
pessimistic.

In the past we have had, partly intentionally and partly by default, a
cheap materials policy, which encouraged the depletion of virgin materials and
the generation of wastes. Now there is increasing interest in conserving virgin
materials and in generating less solid waste. Because recycling works towards
both goals, it has been at the centre of public attention. But recently attention
is turning more to the question of materials policy in general. How does one go
about formulating a materials policy which exploits the flexibility of the
economy, which would be more automatic and effective than appeals to
voluntarism and which would be more conservative in the extraction of virgin
materials and the generation of wastes, at the same time improving the
well-being of both the present and future generations? *Should* we have a more
conservative materials policy?

Material flows

Most people have an idea of what they mean by recycling, but there are
so many different types of material flows in the economy that the concept is a
somewhat ambiguous one. Most would agree that old newspapers saved from
the trash and taken to a paper stock dealer are being recycled. And most would
agree that taking old beverage containers to the store for a deposit return is a
form of recycling. But what about other containers, for example, trucks? Are
they recycled each time they are unloaded, loaded again, and reused? Most

would say so, although there is little essential difference between the two types of containers, except that ownership changes for each bottle reuse while probably not for each truck reuse.

In paper mills some paper ends up on the floor as scrap instead of in the finished product. This scrap, called broke, is normally used as a raw material to make more paper. Most would call the salvaging of broke a form of recycling even though the material doesn't leave the mill and is not bought or sold. A piece of metal that falls off a production line is presumably not recycled when it is picked up and replaced where it fell off. But if it were replaced on the production line at a point before it fell off, would this be recycling?

These questions are somewhat speculative in nature. But when governments try to create incentives to promote recycling they become political and financial. And the point is that recycling is not an easy concept to define with precision.

Given a mandate to increase its purchases of recycled paper, the General Services Administration of the US Government struggled with definition after definition, for over a year, until it hit upon a workable one. It finally defined recycled paper in terms of the material's flow path. Paper flowing from the consumer sector to the economy was 'old scrap'. GSA encouraged recycling by requiring minimum percentages of old scrap content in the paper it purchased. The old scrap could be either discarded or sold. Alternatively the Treasury based its definition on price. Solid waste was material with a zero price. And when the Interstate Commerce Commission was enjoined to hold down freight rate increases on 'recyclables', it based its definition on history. Those materials that were considered recycled scrap in the past become eligible for the hold-down.

These three approaches are to some extent inconsistent. While with careful attention to flow paths and the role of prices it is possible to pin down what is usually meant by 'recycled', there are drawbacks to all working definitions of recycled and scrap, partly because of the changeability of prices and material flow paths. For this reason, legislative efforts to promote recycling which require working definitions of 'recycled' may lead to litigation, unintended effects, and unsuspected windfall gains. But not all efforts to promote recycling depend on a working definition of 'recycled'. For example, the elimination of existing subsidies on virgin materials would promote recycling without requiring new definitions.

Just as prices guide the selection of finished products, they guide the flow of virgin and scrap materials. They determine the scheduling of virgin material extraction from the environment and introduction into the economy and they determine when a material becomes scrap to be thrown away. As long as the price of a material is greater than zero, there is an incentive for the material to stay in someone's hands of ownership. But when no one is willing to pay a positive price for a material and the material itself is a nuisance to the owner, there is an incentive for the owner to disown the material. If there are legal and convenient ways of dumping, fine; otherwise the incentive is to dump surreptitiously – in other words, to litter. Littering, which comes in many forms from beer cans to cars, is the solid waste counterpart of polluting, which gets rid of negatively valued, unwanted liquid and gaseous wastes. In all these cases the problem is the same. It is easy to disown unwanted wastes by dumping them into the environment, imposing costs on others.

Virgin materials and scrap materials substitute for each other and hence compete against each other. The more perfectly they substitute for each other the more important a small change in the relative prices of virgin and scrap materials will be in affecting the amount of recycling. This is true even though material costs are often only a very small part of the total cost of a product.

When Robert McNamara was at the Ford Motor Company, the company made it a practice to buy the latest model Chevrolet and break it down part by part to see how its rival shared costs (Halberstam, 1972, p.231). If they found that a small aluminium casting was doing the job of a small steel part of their Ford car but for a tenth of a cent less, it could mean a total switch in materials for that part and an order for hundreds of thousands of pounds of casting made from scrap aluminium. Obviously a very small difference in the prices of materials can have a very large effect on quantities and types of materials sold with a negligible effect on car prices.

Prices decide how successfully a scrap material competes against a virgin material and whether or not the scrap material will be reused in the economy or discarded into the environment.

Over the past several decades prices have moved in favour of virgin materials and, in a relative sense, against scrap materials, and recycling has declined in many industries. Some of the determinants of price are as follows:

1. Most secondary industries, which process scrap materials are more labour intensive than the primary industries which process the competing virgin materials and which are usually very capital intensive. As labour has become relatively more expensive compared with the cost of capital, the secondary industries have suffered in their competition with the primaries.
2. Technological improvements, which have tended to lower the cost of material processing, have been more concentrated in primary industries than in the secondaries.
3. Compared with secondary industries, the primaries are much more intensive in their use of energy (and have steadily become much more energy intensive). Historically the price of energy has declined, which may have been an important factor in the gradual lowering of costs of extraction in many primary industries over the past hundred years.[1] This too has favoured the primary industries, but now that the price of energy has begun to climb the situation may be reversed.
4. In the past pricing structures have favoured primary industries over secondary ones. In the past energy has been sold on a promotional basis: larger users have gotten substantial discounts. And railroad freight rates have been established on the same promotional basis.

There exists a typical structure of the primary industry and the competing secondary industry for most materials commonly recycled. Most primary industries are dominated by a few large corporations, a situation called oligopoly. Partly as a result of pricing policy in the primary industry, primary prices are slow to move, either up or down. Indeed the posted price for some primary metals remains constant for years. In response to market pressures there are discounts, more or less secret, on the posted price. In this way the actual effective price moves in a range around of the posted price, and the secrecy has confounded attempts to model primary–secondary competition. In contrast there are a large number of secondary producers, each one much smaller and less financially strong than the average primary competitor. Secondary dealers act like commodity traders, which in fact they are. Inventories build and shrink, and quantities cleared in the market are very responsive to price. Prices quoted in the trade journals are very volatile, and there is less of a discrepancy between posted and effective prices.

An important reason for the volatility of secondary prices is the rigidity of primary prices. Without an oligopolistic structure for the primary industry, changes in demand for the final product would be transmitted backward to suppliers of the primary material as well as suppliers of the secondary. But as the price signals are attenuated in the primary industry, the price signals are amplified for the secondary industry. Volatile price movements in the secondary industry encourage a large inventory capacity so that secondary material can quickly be sold or stored. Besides very short-run inventory changes, the secondary industry is more flexible in changing plant capacity and will grow, or shrink, in response to price changes. Because a disproportionate share of market adjustment falls upon them, secondary industries do a favour, of a sort, to the primary industries. They partially insulate the primary industries from the buffets of market fluctuations, making it easier for primary prices to remain sticky.

The role of prices helps explain why voluntary efforts to increase recycling have been largely ineffectual. Prices for scrap materials are usually low, often barely paying the cost of transportation. By adding to the supply of scrap materials, voluntary efforts further depress scrap material prices. Secondary processors find themselves competing with volunteer labour as well as their traditional rivals, the primary firms. Because the volunteer efforts tend to shrink the already thin profit margins of the secondary firms, supplies of material from volunteer efforts tend to displace, rather than augment, supplies from already existing secondary firms.

An economic statement of the belief that there should be more recycling is that virgin materials are too cheap (they are extracted from the environment too cheaply) and scrap materials are not sufficiently valuable. However, so far all we have seen is that scrap prices are low, sufficiently low so that most old scrap is not recycled, and we have seen that prices have moved against, instead of in favour of, scrap materials over the past several decades. We do not really know that prices are 'too' low. In order to be able to tell whether or not we have too little recycling and be able to judge how much material extraction, depletion, throughput, and recycling are just the right amounts, we must have some criteria.

Criteria

By one school of thought it is possible for the market to provide its own criteria to determine the right ('socially optimal') balances of material flows. This view holds that if the market is working properly, there should be no active policies toward recycling or the provision of resources for the future. Whatever amount of recycling or provision the market provides for the future is the 'right amount'. Recycling is not a good in itself, nor is the preservation of resources. When costs from depletion and waste generation fall upon the future, they should be discounted by a market rate of interest just the way other costs, and benefits as well, from market activities are discounted by market rates of interest.

If one feels that somehow the market is recycling too little and not providing sufficiently for the future, he is directed by this view to look for market failures which might explain why the flows are not 'socially optimal'. The three most often cited are the failure of the tax system to be neutral, the failure of the freight rate structure to count adequately the cost of shipment, and the failure of the price system to include the costs of disposal in product price. All three of these failures are considered to favour virgin material extraction over scrap recycling and thus to explain why there is 'too' little recycling and 'too' much depletion. Once these failures are corrected we will automatically achieve the right balance of material flows, which will incidentally imply more recycling and lower rates of depletion than are presently occurring.

While the costs of waste generation are included, conspicuously absent from this list of market failures is the inability of the market to count the costs of depletion which fall upon future generations. It is absent because in this school of thought it is believed that the market does provide adequately for the future. The mechanism is explained by a concept called resource reservation. The owner of a stock of some particular natural resource will decide upon his most profitable rate of extraction by balancing what he can get now for the material and what he may be able to get in the future. As his stock is depletable, he knows that if sold at one time he won't have the stock to sell at another time. Anticipating future scarcity, he reserves some of his stock in order to take advantage of the better market and higher prices associated with future scarcity. Thus material is reserved for future use.

One may wonder if this motivation, which amounts to speculation on future price increases, is not overwhelmed by more immediate considerations. The planning period of the resource holder must be very long, and uncertainties over new technologies and the tenure of resource ownership may lead the present owner to discount heavily (by more than the market rate of interest) the worth to him of possible future price increases. And, in fact, economists who have studied the behaviour of firms in the extractive industries have found little or no evidence of resource reservation.

But lack of evidence of resource reservation may in large part be due to the circumstance that prices of many depletable resources do not appear to have risen over the past century. On the contrary, compared with the general price level, they seem to have fallen, except for timber, and except in the past few years when they have risen dramatically. Some economists have taken this to mean that there is no long-run scarcity problem. New technology has cheapened mining, offsetting the tendency to rising costs from depleting ore stocks, and created substitutes for depleted materials, while new reserves have been discovered. In this view resource reservation is a motivation which has not yet come into play. If natural resource prices were to continue rising, there would be motivation, according to this school of thought, to practise resource reservation and this market incentive would provide adequately for the future.

The above is a very simplified version of the economic theory of natural resources. The policy prescription is to remove market failures. In this school of thought the criterion is to let the perfected market decide. We call this the *efficiency criterion*. The resulting market allocation will depend heavily on the distribution of market power (the distribution of wealth) and discounting, which incorporates the present generation's time preference and partial sense of altruism towards the future.

By a second school of thought, the above is morally indefensible. As decisions concerning the generation of long-lived wastes and depletion of resources affect the means of survival for the future, these decisions should not be made solely by the market in response to the present's appetite and selfishness. Such decisions should be made on ethical grounds.

In the past the ethical problem of how to distribute finite and depletable resources was considered in a simple and uncompromising way. Predictions were made as to when we would run out of iron, oil and other vital stocks. People were enjoined not to waste depletable resources; each should save for the future as a moral duty. This is the ethical problem of the castaway on a barren island with only a depletable stock of hard-tack. He may choose when he will starve and how well he will eat in the beginning, but he cannot avoid starvation. And this is the way that early conservationists formulated the resource depletion problem. The ethical problem was unpleasant as well as explicit.

Fortunately, over the past several decades, the problem of intertemporal distribution has come to be perceived as less unrelenting than this formulation suggests. Deadlines as to when our vital stocks would run out have come and gone. New reserves have been found and new technologies have created substitute materials for old ones. In this sense depletable materials have been kept renewable.

The ethical problem has been transformed without being solved. In the past the costs from depletion were perceived as certain, with certain dates of exhaustion. Now we have begun to perceive the costs as probabilistic. How well can we guarantee that the new technology will come on line in time and that new reserves will be found in time? This same transformation of certainty costs to probabilistic ones occurs on the waste generation side. As we are driven to mine lower grade ores, the amount of waste material and processing increases enormously. What are the risks associated with the increased materials processing and the release into the environment of what were previously trace contaminants such as asbestos, mercury and cadmium? Plutonium is created by breeder reactors. Although it has a toxicity about the same as the Botulina family of organisms, it is scheduled for production in thousands of pounds. Presumably it is handled very carefully, but already it occasionally has leaked into the environment. What risk does this 'solid waste' impose upon future generations?

According to this second school of thought, our choice is whether or not to establish incentives that make it harder to introduce technologies which carry risks such as plutonium, and whether or not to establish incentives that increase the chance of technological improvements which keep us from 'running out'. By cautious policy we could almost certainly guarantee the permanent livability of the planet. Alternatively we could maintain a *laissez-faire* attitude and perhaps find in a hundred years that enough of the risks have turned into realised costs so that we have irrevocably drifted into a hard-tack economy.

In leaving the decision process to the market, even a perfected market, there is no guarantee that the economy will not gradually drift off course, becoming ever more polluted and depleted until it becomes unlivable.[2] And so, according to the second school of thought, the decisions concerning basic resource flows are too important to be left completely to the market, even a perfect one. Some fundamental decisions to guarantee permanent livability should be made explicitly without discounting,[3] but on the grounds of fairness between generations.

These two schools of thought are not so irreconcilable as might be imagined. The theory of resource economics and the theory of resource reservation, so briefly mentioned above (and in the previous readings), is microeconomic theory. This theory asks what is best for a single small mine in the midst of a large industry and whether or not what is best for the individual mine is also best for society as a whole. The analysis proceeds along the assumption that certain conditions hold. The most important for our purposes is that the distribution of wealth is ethically satisfactory and that possible changes contemplated are small enough not to effect the distribution of wealth in a socially undesirable way. This along with other conditions such as full employment and no inflation are macroeconomic conditions and they form an environment in which microeconomic firms and markets operate.

What is good for one particular mine, however, may not be good from society's point of view when all the mines are acting together. For example, one particular mine can become depleted in the midst of an industry with ample reserves without general depletion of the mineral, but if all the mines become depleted at the same time then the mineral is indeed depleted. The theory of exhaustible resources was designed to analyse cases of the former variety where the important macroeconomic conditions are stable. But in the latter case, where an entire mineral stock is depleted, or where some long-lived waste is generated, the distribution of economic well-being can be enormously affected. The macroeconomic environment can no longer be taken for granted, and indeed the macro environment becomes the focus of study.

The second school of thought is concerned directly with the macro environment. The ethical problem raised by this school can be illustrated in an economic context by comparing the market allocation of resources over time with the market allocation of resources at a given moment of time. It is true that the market provides a well-defined and explicit distribution of goods and services at a given moment of time. But is this distribution ethically desirable? This question immediately invites another question: 'is the distribution of market power socially desirable?' If this distribution of wealth, and hence market power, is found in some way ethically undesirable then the market allocation stemming from this distribution of wealth is also ethically unsatisfactory. It is well known in economic theory that the ethical question of the distribution of economic well-being is logically prior to the market allocation of goods and services.

What does this observation imply about the intertemporal allocation of resources, in particular the flows of depletable resources and the generation of long-lived wastes? For us to be ethically satisfied with the market allocation of flows we would have to be ethically satisfied with the implicit distribution of market power – that is, the distribution of wealth intertemporally. By the most usual interpretation in economic theory, the intertemporal distribution of wealth (and hence market power) is that all resources belong to the present and none to the future. To the extent that the future is provided for, it is provided out of a sense of altruism by the present. What we have is a consumer sovereignty of the present generation, a sovereignty over all future generations. While it presumably benefits the present, many people living in the present may find this *de facto* arrangement ethically unsatisfactory.

One can see the difference between concepts of altruism and fairness by comparing the resource allocation problem intergenerationally with another resource allocation problem which is occurring across groups of people at the present moment of time. Many people think that the ocean's resources are owned jointly by all people in the present generation. One rule of fairness would be to establish compacts among the fishing nations to ensure that each one got an equitable catch and the stock as a whole was not overfished. Another way of allocating the fisheries resources would be to give control and full rights for the entire resource stock to Japan for the first year, to the US the second year, to Russia the third year, and so on. Management of the stock would depend on the 'present generation's' inability to exhaust the resource in a single year and upon its altruism toward the 'next generation'. Relying on the altruism of previous nations would doubtless make successor nations feel uneasy. Yet this is roughly the situation with natural resources intertemporally.

Thus we can identify the first school of thought with the microeconomic allocation of the market and the second school of thought with the macroeconomic concern for a fair or just distribution of economic well-being over time. The two schools of thought are not in conflict; in fact, they take place on different levels. We can compare the conservationists' goal of permanent livability with other macroeconomic goals such as the control of inflation and unemployment. The question is not economic efficiency or something else, but what kind of environment is best for the market system (perfected as much as possible) to exist in. For example, we may decide that an 8 per cent inflation is 'unacceptable', but a 3 per cent rate is 'acceptable'. The money supply and other policy tools are then manipulated to achieve the macroeconomic goal. Within the environment set by the macroeconomic policy tools the market system is expected to operate freely as before, but better tuned to broad social goals. Similarly macroeconomic policy tools can be manipulated to guarantee more nearly the goal of permanent livability.

In the case of income distribution intratemporally, the poor and the rich are present at the same time and there are political mechanisms to move towards what is considered a fair distribution. The poor and the rich can both be represented legislatively (the fairness of the representation is another matter) and a progressive income tax established. But the future is not around to press any claims against the present, and potentially the present does have all the market power.

Granting that there are people, in the present, who wish to see provision for the future based on a fair distribution of well-being rather than on a consumer sovereignty of the present, with no actual representation by the future, how are these people to define what is fair and how are they to convince sufficient numbers of others in the present to go along with the implied policy changes? One way to develop ideas about fairness intertemporally is to undertake the following thought experiment. Imagine a constitutional convention of delegates chosen across time. Under a 'veil of ignorance' these delegates do not know which generation they might be born into and they are charged with the task of agreeing on rules of fairness, rules which are acceptable from any vantage point in time.

Under these conditions it is easy to imagine that the goal of permanent livability would be very attractive. While there might be considerable controversy over the best means to guarantee this end, one rule of thumb recommends itself: let each generation be self-sufficient. As one generation depletes a resource it renews the resource by means of a better technology or replaces it with another just as good. At the same time no generation burdens the future with some pesticide or whatever that leads to a cancer epidemic a generation later. In this way no generation borrows from a later one and generation will follow generation indefinitely. (If the policy steps enhancing the chances of permanent livability are inexpensive, they are likely to be attractive to the present generation without appeals to ethics and social contracts.) Another way of putting the matter is that the earth as a whole should be treated as a sustainable yield asset. This idea is labelled the *conservation criterion*.

The discussion of criteria takes place on two levels. On the micro level there are particular firms and particular markets, and the question is how well, or efficiently, does this micro system function, assuming that it is embedded in a satisfactory macro environment. On the macro level we ask whether the entire micro system of individual firms and markets is drifting off course and what can be done to keep it on course. On the micro level policy alternatives may be evaluated by the efficiency criterion and on the macro level by the conservation criterion.

Market failures

The microeconomic system of material flows can be considered in more detail by focusing on the three areas of alleged market failure. This differs from the usual positive approach to this problem in that this time there is a normative standard to measure against the (micro) efficiency of markets. While the picture that emerges is by no means clear, it does appear that in terms of the micro system itself there are indeed market failures leading to too much material throughput and not enough recycling.

Freight rates and volume discounts

There are many claims that freight rates in the United States discriminate against scrap materials in favour of virgin materials. In support of these claims, comparisons are brought forward showing rates two and three times higher for the scrap materials than the competing virgin material counterparts. Because freight rates are a large fraction of the value of many scrap materials, discriminatory rates would indeed adversely affect recycling and promote virgin material extraction. However, the railroads and the Interstate Commerce Commission have counterargued that the comparisons brought forward are isolated cases and are thus unrepresentative and that the difference in rates, where they occur, are justified by the differences in the costs of shipment.

Unfortunately, this controversy is far from resolution. There are reputed to be 43 billion rates; often there is a special rate for each commodity moving from each origin to each destination. The rate structure has grown up in a haphazard way over the years in response to historical pressures. From this morass economists have tried to select a few overriding factors which can be taken as organising principles to make sense of the rate structure. And under increasing pressure in the last decade, the ICC has begun its own study of the rate structure in an effort to base it more on cost realities and to modernise it. Environmental groups have challenged the ICC in the courts. In response the ICC has prepared an environmental impact statement on scrap recycling and virgin material shipments. And in the past few years freight rates have become more favourable to scrap shipments, relative to virgin materials.

Although the ICC has not paid much attention to economic analysis in the past, its study, Ex Parte 270, could be very helpful in settling the question of freight rate discrimination. In the meantime we are left with 'organising principles' and fragmentary, untrustworthy data. One of the organising principles is demand pricing or what the railroads call value-of-service pricing. By this principle commodities which are less sensitive to price increases bear higher freight charges. For cheap bulky materials the freight charge is a large fraction of the materials' total value and whether or not these materials can afford to travel at all depends sensitively upon the rate charged. (The same principle of demand pricing is also important for natural gas and electricity. Those customers that are most sensitive to price increases are favoured with the lowest prices. The most sensitive customers turn out to be the largest customers, and for energy the principle of demand pricing leads to volume discounts for the large users.)

The ICC's cost studies, called burden studies, seem to bear out this principle of demand pricing. Both scrap and virgin materials, which are cheap and bulky and sensitive to price changes, appear to be subsidised relative to the costs of shipment and the rates for other commodities. Thus the rate structure encourages too much throughput by incentives on both ends of the economy. In addition, it seems from these studies that scrap materials are considerably less favoured than virgin materials. Reversing Pangloss, we might say that for the conservationist this is the worst of the worst possible of worlds; first to subsidise material throughput at both ends of the economy and then to retard recycling by subsidising virgin materials more.

Tax provisions

The percentage depletion allowance[4] is probably the best known subsidy to extractive materials. Although it shelters more income from taxation in the oil industry than for all the other extractive industries, the oil industry is also bigger than the other extractive industries combined. Thus the depletion allowance is likely to be just as important relative to other extractive industries as it is to oil. The depletion allowance was originally intended to be a temporary measure to smooth out one of the inequities of the imposition of the income tax in 1913. When the income tax came into being, mine owners who had realised their income before 1913 were in a better tax position than ones who had developed their mines but had not yet realised their income by 1913. The depletion allowance was designed to allow mine owners to realise tax-free income which had accrued before 1913. In a few years, many thought, the allowance would become moot. On the contrary, it has prospered and now its modern form, the percentage depletion allowance, subsidised not just gas and oil but practically all extractive resources (the big exception is timber). Basically the allowance functions like a negative sales tax, cheapening the price of materials for consumers and enlarging the profits for producers. The profits tend to be competed away as the extractive industries are enlarged, but in the process more material throughput is encouraged.

The depletion allowance is by no means the only important tax preference. Virgin material industries are also benefited by capital gains provisions, foreign tax credits, and special property tax arrangements. These are partially offset by other special taxes on the extractive industries, such as state severance taxes. Some of these favourable tax provisions of course affect more than just the extractive industries, but they favour the extractive industries more than other industries. In consequence, the extractive sector is one of the most lightly taxed sectors in the economy. Compared with a neutral tax system, the present tax structure misallocates resources into the extractive industries, increasing the rate of extraction and favouring virgin material industries over scrap industries.

The failure of the market system to include the cost of disposal in the product price.

The cost of disposal includes the municipal costs of waste collection and processing, the residual costs during and after processing (air pollution from incinerators and the like), and the aesthetic and other economic costs of wastes not collected (litter). For many products a century ago the costs associated with disposal were a small fraction of a product's price, perhaps a per cent or so. Many waste products were of a natural origin and decomposed rapidly. There were many handy sites, where wastes could be thrown at negligible cost to society. Now the national average cost of collection and processing, just a part of the total disposal cost, is somewhere around $20 a ton,[5] and in congested areas it is much higher. The cost of littering to society, on a per unit basis, is considerably higher than the cost of treating wastes deposited into collection streams. And as products have grown more complicated in composition, they sometimes contain more materials which are toxic and dangerous after disposal. Nowadays in many cases the costs associated with disposal are a much higher fraction of the product's price. When costs associated with disposal grow to 10 per cent or more of the product's price, there is a serious distortion in the price system. Material-intensive activities are favoured too much compared with less material-intensive activities.

Municipal solid waste disposal is 'free' from the point of view of the disposer. For a particular item disposed, the disposer's tax payment for waste management is diluted by everyone else's tax payment. One's tax payment is a negligible function of one's own disposal. Emissions of gaseous or liquid pollutants is a method of 'free' disposal from the point of view of the polluter. Both uncontrolled pollution emissions and 'free' solid waste disposal, even though encouraged and sanctioned by the state, constitute failures of the price system to internalise costs. (A cost is internal when the cost is borne by its generator.) The latter case of market failure, the failure of the price system to include the cost of disposal in the product price, is a general problem directly relating to solid waste generation, scrap markets, and the 'overly' inexpensive virgin materials. But it also should be stressed that the failure of the market system to include solid waste disposal costs is not different in principle from the failure of the market system to include pollution costs from liquid or gaseous wastes.

Over the past decade there have been increasing efforts to control the cost of polluting discharges. It turns out that compared with the rest of the economy, both virgin material and scrap industries are relatively polluting. The implication of this is that for general abatement throughout the economy basic materials, both from scrap and virgin sources, will become relatively more expensive compared with the general price level, so that there will be increased incentives for products of greater durability and lower material intensiveness.

Clearly if both a virgin material industry and its competing scrap industry are required to abate their generation of pollution, the relatively cleaner industry will be at an advantage compared with its competitor. While both are relatively polluting compared with the economy in general, it appears that for several basic materials the primary industry is more polluting than the competing secondary industry.[6] Part of the reason for this is that the processing chain is longer for the primary industries, extending all the way back to the mine mouth. This would indicate that pollution control will favour scrap flows compared with virgin material ones.

Policy

As we have seen, discussion of criteria can take place on two levels, and correspondingly policy prescription should take place on two levels as well. On the micro level of individual firms and markets it is appropriate to prescribe policy to eliminate the market failures. For the formulation of a materials policy this primarily means prescribing policy to eliminate the failures. In principle this is a well-defined problem. But in practice it is not cut and dried; some of the market failures were established to serve some social purpose and trade-offs must be made. None the less this is the kind of work economists are accustomed to undertaking.

On the macro level it is appropriate to prescribe policy to ensure that the entire micro system of markets and firms does not drift off onto some undesirable course. This type of policy making is considerably more elusive. It is very hard to tell whether or not we are drifting off course, let alone to tell just how large a correction is needed. Before we can be confident as to the appropriate corrective dosage we need to know much more about reserves, environmental costs associated with extraction and throughput, new technologies, and future population and consumption patterns. These are tough subjects and we are unlikely to learn enough to be precise in our prescription. Still, some of the medicine is relatively painless to swallow and it is much less harmful to err a little on the side of caution than to err a little on the side of indulgence. In formulating macro policy for the large flows of materials, it needs to be remembered that the medicine may work exceedingly slowly, as the problems (and solutions) lumber along with great momentum, but that some of the medicine may be quite cheap.

Notes

1 Data from Potter, N. and Christy, F. (1962) *Trends in Natural Resource Commodities* (Johns Hopkins Press, Baltimore), suggest that this factor may not be too important, see pp. 33–6.
2 It is sometimes carelessly thought otherwise.
3 'Without discounting' does not mean discounting at a zero rate.
4 'Depletion allowance' and 'depletion deduction' refer to the same tax concept. The depletion 'allowance' is just a deduction from taxable income and so the latter term is used from now on.

5 *The Environmental Protection Agency* estimated $18/ton for 1974. See *Second Report to Congress: Resource Recovery and Source Reduction*, Office of Solid Waste Management Programs, USEPA, 1974, p. 6.
6 The EPA compares the pollution costs associated with several virgin and scrap materials in the *First Report to Congress: Resource Recovery and Source Reduction*, Office of Solid Waste Management Programs, USEPA, 1974, pp. 1–9.

References

Halberstam, David (1972) *The Best and the Brightest*, Random House, New York.

Chapter 6
Social welfare and exhaustible resources
Paul Grout

The questions economists are expected to answer concerning exhaustible resources tend to be of the following two types. Firstly, how is the stock of exhaustible resources likely to be depleted over time and how will this change as rates of interest and prices are changed? Secondly, is this depletion path too fast, too slow or, more specifically, what is the optimal rate of depletion of the exhaustible resources? The latter questions fall into the field of economics traditionally referred to as welfare economics and it is with these questions that this chapter will be concerned. The aim is to highlight the possible concepts of social welfare which could determine the optimal rate of depletion of exhaustible resources and to discuss their advantages and disadvantages as concepts of justice. A simple theoretic model is used which requires little knowledge of economics or mathematics to follow.

To say that society has too high a rate of resource depletion implies social welfare would be higher with a slower rate of depletion which in its turn implies that a concept of social welfare exists which allows us to compare possible alternatives. Students will be well aware from their basic economics courses that any economic statement that one state of the world is better than another involves value judgements, and will also be aware from everyday conversation that all individuals do not share the same value judgements about what is good for society and what is bad. The immediate implication of this is that in general there will be no rate of depletion which everyone agrees is the best possible. An economist can only state that a particular rate of depletion is best given a specific concept of social welfare. Thus the relevant discussion of what constitutes the optimal rate of depletion is really a discussion of what constitutes social welfare (assuming all technical parameters are known).

The role of the economist in the determination of social welfare and thus the role of welfare economics within the subject is itself a point of discussion. One school of thought rejects welfare economics as a valid subject. Reasons frequently used to defend this view are: firstly, that by definition all concepts of welfare are based on value judgements and that economics should attempt to avoid all value judgements in its analysis; and, secondly, that when final decisions are taken political short term views tend to be paramount and concepts of social welfare as utilised by economists are of no relevance. A strong advocate of the irrelevance of welfare economics as a subject of study is

Graaf (1957) which Baumol (1952) described as bearing an 'ill-concealed resemblance to obituary notices'. The problem with the first of these reasons seems to be that someone has to make the value judgements, resources have to be used and economists should not ignore their essential role in the final choice. Even if economists do not wish to directly enter the debate by throwing their weight behind specific welfare concepts their assistance is still required in analysing the economic consequences of alternative value judgements, a role economists have been increasingly called upon to fill in the last decade, particularly with regard to exhaustible resources. As long as the economist makes clear what value judgements he is incorporating into his results the study of welfare economics is a crucial part of the economist's contribution to the analysis of whether present rates of depletion are optimal or not.

Concerning the second argument one can agree that the way politicians treat today's generation compared to the next may be influenced by the fact that today's generation have today's votes. However even with this cynical view there may still be many decisions where the politician may be willing to utilise less personal concepts at welfare. Various aspects of the depletion of exhaustible resources will fit into this area, e.g. if a politician must make a decision which will affect the relevance of resources for the next century there seems no reason for his concern for votes and interparty politics to influence his view of how to treat the generation alive thirty years hence in comparison to the generation fifty years hence. Furthermore, given a particular rate of depletion advocated by a group or political party in society, the implicit assumptions can be considered by the economist, such consideration increasing the internal consistency in the groups views and persuading them to change their stance if necessary. At least the economist's analysis would allow alternative groups to focus on the basic assumptions of the initial suggestion which should generate a more consistent and accurate discussion of the points of disagreement. With this reasoning in mind we will consider the ethical foundations and implications of the most common views of what constitutes society's welfare.

There is clearly a trade-off between clarifying the ethical foundations of alternative views and being accurate in predicting their implications for the depletion of exhaustible resources. The former is laid bare most easily in simple models, while the latter is satisfied by introducing as many of the technical parameters as possible (e.g. the cost of exploration, the costs of alternative techniques of extraction, how uncertain are we about innovations). We will concentrate on a simple model which catches the essence of the problem and introduce extensions which help in clarifying issues rather than predicting what the specific rate of depletion should be in a particular society.

The simplest model we could think of is a discrete-time model, production in each period requiring a scarce resource (R_i is amount of resource in period i), the total quantity of which is fixed. Given that the amount produced in each period is increasing with the quantity of resource used one could further simplify by assuming each unit of resource (approximately measured) generates one unit of consumption, i.e. constant returns to scale in each period. If the total amount of resource is $\bar{R}$ and we have equal consumption of the resource in each period, then given m periods the resource available per period must be less than or equal to $\bar{R}/m$. When the number of time periods becomes infinite the amount of resource per period is

$$\lim_{m \to \infty} \frac{\bar{R}}{m} = 0$$

Thus each time period has no resource. Now no matter how large or small we pick our time periods the number of time periods will be infinite, stretching off forever into the future. This gives us an important result concerning exhaustible resources which is that if the total amount of resources in society is fixed then it is only possible to give every time period an equal share of the resource if each period gets none. However such a solution is a blatant waste of resources; thus, if we do not wish to waste scarce resources, then as long as infinite production is not possible with a finite resource an equal distribution is out of the question. In fact, the amount of resource available to each period must clearly decline over time otherwise the condition

$$\sum_{i=1}^{\infty} R_i \leqslant \bar{R}$$

would not hold. While this in itself is an interesting property, fortunately very few people believe that society will have to rely on the resources we use at present to generate consumption for ever more. It seems reasonable to assume that new techniques will be found to produce energy which, while perhaps still requiring resources which are exhaustible, will not require the resources we are presently concerned with. While the date of implementation of such inventions is obviously uncertain such uncertainty will only confuse the analysis and so we shall assume there exists a period n after which consumption can be produced at a constant amount per period, $\bar{c}$, and that we have an exhaustible resource which is essential to production in all periods from today, 1 to n. The constraint society now faces is

$$\sum_{i=1}^{n} R_i \leqslant \bar{R}$$

The time periods being used are best interpreted as being long, so that we can think of each period as being one generation, the next their children, the next their grandchildren, and so on. Interpreted this way the concepts of welfare seems to catch the essence of the conflict. Whether we are making a mistake concerning consumption rates today with two years hence is an important question, but this is less worrying and we are less likely to be wrong than in the question of whether we are making a mistake concerning consumption rates between this generation and all the following generations for the next fifty to a hundred years. Alternatively, if one wishes to conceive of the time periods as being shorter or longer length then this will not affect the analysis.

Having defined each period as a generation, one then has to represent the preferences of each generation. Given the simple form of production that we have assumed one can define utility on the resource and any increasing function will suffice, since all we need is $U_i(R_1) > U_i(R_2)$ if and only if $R_1 > R_2$. If one wishes to go further and, rather than just represent preferences, actually attach welfare significance to the utility levels, e.g. saying generation one is better off than generation two, $U_1(R_1) > U_2(R_2)$ then one has to be more careful. It is frequently observed that utility functions exhibit diminishing marginal utility $[U''(R) < 0]$ and this will be assumed throughout. A reasonable starting point will be the assumption that all generations are basically similiar and therefore all their utility functions will be identical. It must be emphasised that such a statement is again a value judgement since at the present time there is no known method of proving that one individual or group of individuals are as well off as another group of individuals in welfare terms, although one may feel confident that observation of individuals tells one a great deal about their welfare. With these assumptions and setting $n = 2$ the possible alternatives for society can be seen in Fig. 6.1(a) showing alternative distributions of the resource, (b) giving the utility equivalent, the shape of the feasible set being the result of diminishing marginal utility. In order to obtain some insight into the various concepts of social welfare this case and two other extensions will be considered.

The first extension concerns the possibility of economies of scale in extraction of the resource in each period. This implies that if we extracted the whole resource in one period there would be more in total than if we extracted on equal quantity in each of the n periods. Given reasonable assumptions the results shown in Fig. 6.1 would change and take the form of those in Fig. 6.2. Note that if 6.2(a) is symmetric around OA (implying the same extraction technology is available in each period) then 6.2(b) will be symmetric around OB. While the shape of 6.2(a) is obvious the shape of 6.2(b) may be less clear.

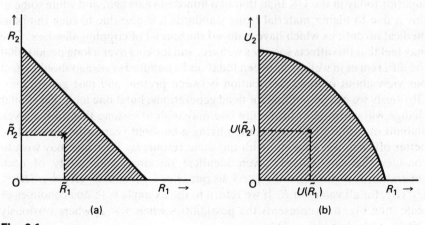

(a)   (b)

Fig. 6.1

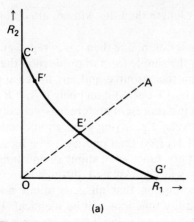

(a)

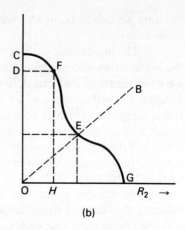

(b)

Fig. 6.2

For each small movement from the point C to F in 6.2(b) the increase in utility to generation One is greater than the loss to generation Two even though the quantity of resource given up by Two is greater than that gained by One. The reason is that since generation Two is relatively rich in the resource compared to generation One the marginal utility of generation One is sufficiently higher than that of Two to offset the variation in resource as we move from C' to F'. Obviously, due to diminishing marginal utility, this differential in marginal utility is falling and after the point F (i.e. from F' to E') the differential is no longer sufficient to offset the increasing returns to scale. Moving from the point E' to G' uses an identical argument with the differential in marginal utility being reversed.

The second extension concerns the assumption of the utility functions of each generation. In the short term it is reasonable to assume successive generations will be fairly similar but over a long period this may change. To give an example, the state of health of the majority of the present generation is superior today in the UK than that of a hundred years ago, and while some of this is due to higher material living standards it is also due to such things as medical inventions which have reduced the spread of crippling diseases. One may feel that this affects society's welfare, and looking over a long period, that the differences in welfare between today and a hundred years ago should affect our view about optimal distribution between present and past generations. Obviously we can do nothing for dead generations, but if one feels that similar change will carry on in the future one may wish to assume that with a given amount of resources the generation living a hundred years from now will be better off than we would be with the same resources. Thus one may wish to consider the case where, given identical resources, the utility of each generation is higher than the previous one, i.e. $U_1(\bar{R}) < U_2(\bar{R})$, $U_1{}'(\bar{R}) < U_2{}'(\bar{R})$, for all values of $\bar{R}$. If we return to the example with no economies of scale then Fig. 6.3 represents the possibilities when $n = 2$, where obviously OB' = OA' but OB > OA.

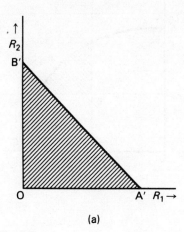 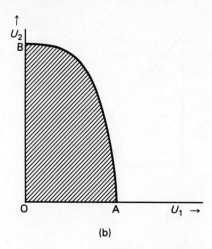

(a)   (b)

Fig. 6.3

Given that any statement such as 'we are depleting our resources too slowly or quickly' requires a value judgement, it seems reasonable to try to find value judgements which most people will find acceptable. Consequently it seems sensible to start with the concept of welfare which has been least criticised in economics, that of Pareto efficiency. A distribution is Pareto-efficient if there is no other distribution which allows everyone to be as well off and at least one individual to be better off (this is sometimes referred to as strong Pareto efficiency). Specifically within our model we define a distribution of the resource $(R_1 \ldots R_n)$ as Pareto-efficient if there exists no alternative distribution $(\bar{R}_1 \ldots \bar{R}_n)$ satisfying $\sum\limits_{i=1}^{n} \bar{R}_i \leqslant \bar{R}$ with

$$U_i(\bar{R}_i) \geqslant U_i(R_i) \text{ for all } i = 1 \ldots n$$

and

$$U_i(\bar{R}_i) > U_i(R_i) \text{ for at least one } i.$$

Thus a distribution will only be Pareto-efficient if it utilises all of the resource in the n time periods, since if it did not, e.g. A', B' in Fig. 6.4, then there would exist distributions where both generations One and Two are better off, and of course all other generations get $\bar{c}$, hence the distribution could not be Pareto-efficient. The concept of Pareto efficiency allows us to reject an infinite number of possible depletion paths, but since all paths satisfying $\sum\limits_{i=1}^{n} R_i = \bar{R}$ are Pareto efficient (all points on the utility locus CD in Fig. 6.4(b)) the problem of whether we are depleting too fast or too slowly cannot be answered. Historically the term Pareto optimality has been used instead of Pareto efficiency: however, it seems strange to label all the points on the locus CD in Fig. 6.4(b) as optimal since some imply rapid depletion of resources and others imply no

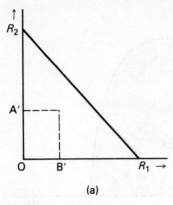

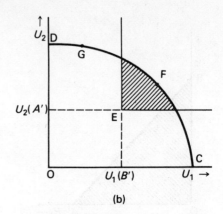

(a)   (b)

Fig. 6.4

depletion of resources. It seems most sensible to consider this concept as an efficiency condition which reduces the set of possible time paths we need to consider but cannot tell us which of this smaller set is optimal. One may be surprised to find that even this very weak concept does not meet with full acceptance. Some economists, e.g. Peacock and Rowley (1975), are willing to accept a society which uses its resources inefficiently if the only alternative impinges on individual freedom. There are others who would criticise the concept for paternalistic reasons. An individual may spend 100 per cent of his income on beer but others may believe society would be better off if he should drink less and thus would favour constraining his drinking habits, making him worse off even though no one is made better off in the process. The rejection of the Pareto criterion by some economists makes it very clear that acceptance of the Pareto criterion is based on a value judgement. This is a point frequently forgotten in economics, textbooks and economists often giving the impression that efficiency as we have defined it is a basic ingredient of the subject.

Not only does the Pareto criterion fail to choose between an infinite number of depletion paths but also it does not allow us to rank all Pareto-efficient paths as preferable to non-Pareto-efficient paths. In Fig. 6.4(b) point F is clearly preferred to point E but the Pareto criterion tells us nothing about the relationship between E and G even though G is Pareto-efficient and E is not. To make any real statement about the depletion of exhaustible resources we will need to make stronger value judgements and it is really from here on that most disagreement is met.

An attempt to extend the ranking of states of the economy beyond that implied by Pareto efficiency, but still to remain close to the concept, was made by Kaldor and Hicks (see Graaf, 1957, for discussion) who introduced the concept of compensation criteria. The basis of their contribution is hypothetical lump-sum compensation, i.e. compensation which does not take place but is only theoretically considered and is of the form that the gainers give up a fixed quantity of output which is received by the losers without any transactions costs and without any effect on production. Specifically within this model we can say that depletion path $(R_1, \ldots, R_n)$ is preferred to $(\tilde{R}_1, \ldots, \tilde{R}_n)$ if given a move from the latter to the former the gainers can compensate the losers and still be better off. It is important to remember that compensation is not paid since if compensation were forced to take place then the rule would simply reduce to Pareto efficiency. In terms of our basic model without economies of scale in extraction the compensation criteria would make all points on the locus CD in Fig. 6.4(b) acceptable since there would be no depletion rates which are superior using the compensation criteria. However if we introduce economies of scale in extraction the picture changes dramatically as Fig. 6.5 explains.

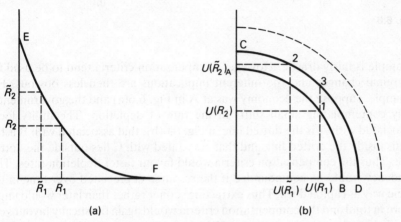

Fig. 6.5

The locus AB shows the theoretical utility levels for generations One and Two as lump sum redistributions take place, given the economy is at (R_1, R_2), e.g. the point B shows utility of generation One if generation Two pay a lump-sum compensation to generation One of R_2, at the other extreme A shows utility of generation Two if generation One pay them a lump-sum transfer of R_1. Note that production (or in this case extraction) is held constant at (R_1, R_2) for these compensations. Alternatively, suppose the economy is at $(\tilde{R}_1, \tilde{R}_2)$ then the theoretical utilities which can be reached are shown by the locus CD. Since CD lies completely outside AB we can say $(\tilde{R}_1 \tilde{R}_2)$ is preferred to (R_1, R_2) by the compensation criteria. The reason for this statement is that if the economy shifted from (R_1, R_2) to $(\tilde{R}_1, \tilde{R}_2)$, from 1 to 2 in Fig. 6.5(b), then the gainers could compensate the losers (i.e. move to point 3 where generation One have the same utility as at 1 and are still better off (generation Two have higher utility at 3 than at 1). The implications of compensation criteria are clear from Fig. 6.5(a) and (b). The optimal distribution of resources is either E or F in Fig. 6.5(a) since the utility locus associated with each of these is the dotted line Fig. 6.5(b) which lies outside every utility locus associated with all points in Fig. 6.5(a) between E and F. The implication for resource depletion in this

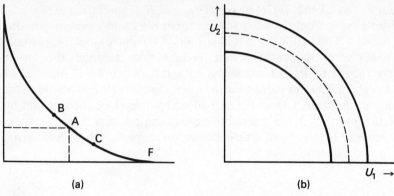

(a)   (b)

Fig. 6.6

example is fairly drastic. However, compensation criteria tend to be used for marginal changes and the inherent implications are then less obvious. For example, suppose the economy was at A in Fig. 6.6(a) and the government is only contemplating small shifts in the rate of depletion. The utility locus associated with A is the dotted line in Fig. 6.6(b), that associated with B is the locus inside the dotted line and that associated with C lies outside the dotted line. Thus the compensation criteria would favour faster depletion rates. This tendency would be compounded if the resource depreciated over time in the same way as capital does. Thus extracting sooner rather than later would imply more in total and the compensation criteria would again drastically favour very rapid depletion of the resource. This favouring of faster depletion rates is independent of the level of utility of each generation, e.g. in Fig. 6.6(a) the preference for C over B would still hold no matter how close the three points were to F. This isolates the failing of compensation tests, namely that they do not take account of the actual levels of income but only what would be given in a hypothetical transfer, thus ignoring the problem we are most concerned with. It is little consolation to tell a poor man that a hypothetical lump-sum transfer would make him better off at the same time as one reduces his income still further. It appears that compensation tests are a poor representation of social welfare when we are considering fundamental changes to the economy such as altering depletion rates of exhaustible resources.

   A consumer planning his consumption over his lifetime is often assumed to maximise his total lifetime utility. Since we have utility functions defined over each generation's consumption one could adopt the similar approach to society's welfare and choose the time path which maximised the sum of all generations' utility. This approach is referred to as a utilitarian welfare function and is the most historically entrenched view of social welfare; traditionally considered as needing no justification since it was taken as self-evident that society should be ordered to create 'the greatest happiness for the greatest number'. It was not until the 1930s that the almost exclusive use of the utilitarian philosophy in economics was criticised (see Robbins, 1938). One may notice that in the specific model being used, since all generations beyond n received $\bar{c}$, the sum of utilities must be greater than or equal to

$$\lim_{q \to \infty} q \; U(\bar{c}) = + \infty$$

so that all uses of our resource will give the same total utility. However, if we stop at any q, after n, the utility sums from 1 to q will depend upon how the resource is distributed and the optimum utilitarian solution is to distribute the resource to maximise the utility sum at any $q > n$.

Obviously it would be helpful to try to justify the adaption of such a rule and one of the most impressive attempts to do this has come from Harsanyi (1955). To understand this argument we must first discuss how individuals evaluate levels of income which are uncertain. If an individual gets income y, for certain then his evaluation of it is $U(y_1)$. If an individual may get income y. with probability, p_1, and y_2 with probability p_2, $p_1 + p_2 = 1$, then it is argued that a rational individual will evaluate this by

$$p_1 U(y_1) + p_2 U(y_2)$$

i.e. how much y_1 is worth to him times the likelihood that it will occur, plus how much y_2 is worth times its likelihood to occur. Indeed it has been shown that given three 'axioms of rationality' individuals will always evaluate uncertain returns in this way. Now one can argue that the reason individuals do not agree on what is, say, a fair rate of tax is that the discussion of such things takes place in the full knowledge of how rich or poor one is. Nobody expects a millionaire to agree with an unemployed person on what constitutes the correct tax to be paid by the rich to redistribute to the poor. The millionaire knows he will be paying a large amount of any tax required and will consequently favour lower tax rates while the person who is unemployed will know he will receive tax (a subsidy) and will tend to favour high rates of tax. Suppose, however, they did not know which of them was the millionaire and which was penniless: surely then their attitude would change. To take an extreme example, suppose both were unemployed and both were possible beneficiaries of a will made by a now deceased millionaire, the will stating that one of the people was to receive a million pounds, the other nothing, and the decision to be made by the toss of a coin. If the two individuals were offered the alternative of receiving half a million pounds for certain they would surely accept, and thus both would have agreed on the optimal distribution. Harsanyi argues that one's ethical preferences should be independent of one's position in society; if one's views change depending on whether one is rich or poor then these preferences must be selfish rather than ethical in Harsanyi's sense. He then argues that an ethical preference will therefore not be revealed in choices made when one knows whether one is rich or poor. He believes that ethical choices would only be made if one 'did not know one's position in society . . . but rather had an equal chance of obtaining any of the social positions existing . . . from the highest to the lowest'. It is clearly impossible to put someone in this position, but the final thread of Harsanyi's argument is now clear. If a 'rational' individual's valuations over uncertain outcomes can be predicted as we have shown above then surely if ethical preferences are 'rational' we can make similar predictions. Thus, as long as we are willing to accept that the ethical preferences must be 'rational' the evaluation of the possibility of being placed in the position of each of the n generations is

$$\frac{1}{n}U_1(R_1) + \frac{1}{n}U_2(R_2) + \dots\dots\dots\dots\dots\dots\dots +\frac{1}{n}U_n(R_n)$$

which is always maximised when

$$U_1(R_1) + U_2(R_2) + \dots\dots\dots\dots\dots\dots\dots +U_n(R_n)$$

is maximised, the latter being the utilitarian concept of social welfare.

In the example when $n = 2$ one can draw indifference curves in utility space, one for each K,

$$U_1 + U_2 = K$$

which shows combinations of U_1 and U_2 having the same social welfare value, K. These indifference curves will be straight lines and are shown in Fig. 6.7.

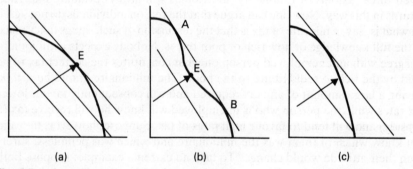

(a)   (b)   (c)

Fig. 6.7

Figure 6.7(a) shows the simple case without any economies of scale in extraction, the optimal depletion path being a constant quantity in each period. Figure 6.7(b) gives the economies of scale in extraction, either A or B being an optimum extraction rate, both of which generate more total utility than the equal quantity per period situation. Figure 6.7(c) shows the solution of $U_1(\bar{R})$ $< U_2(\bar{R})$, $U_1'(\bar{R}) < U_2'(\bar{R})$ for all $\bar{R}$. The important point about this final example is that more of the resource goes to the generation who have the highest utility for any given level of income. This property may run counter to many people's intuitive beliefs about justice. For example, consider the thalidomide children who, given a strict utilitarian concept of welfare, would not only *not* be given special consideration because of their position but would actually be given a smaller level of income than normal, healthy children. Individuals who can benefit most from society's production are given the largest share since they can contribute most the 'greatest happiness for the greatest number'. Thus if we feel that future generations will have better technical knowledge and thus produce more goods from a given quantity of resource than we can at present, then the utilitarian rule would dictate that we reduce our consumption of the resource and increase the quantity that we pass forward to future generations. The more productive generations are therefore better off for two reasons, firstly because if the resource was equally distributed they could produce more than the less productive generations, and secondly because they will actually receive more of the resource than is allocated to the less productive generations.

If one wishes to disagree with the utilitarian philosophy then one could criticise Harsanyi's general approach or more specifically disagree with the equal probabilities assumption (this will be discussed later) or with the imposition of 'rationality' on social preferences. This latter criticism has tended to focus on one specific condition which the so-called rational preferences must satisfy, and it is argued that one's ethical preferences may conflict with this condition. This specific condition states that if we can identify two alternatives (say α and β) which an individual is indifferent between ($U_\alpha = U_\beta$), then the individual must be indifferent between the following two alternatives, $\bar{\alpha}$ and $\bar{\beta}$ $\bar{\alpha}$ being that α occurs with probability 0.5 and δ occurs otherwise, and $\bar{\beta}$ being that β occurs with probability 0.5 and δ otherwise, for all δ (i.e. for all possible alternatives). If an individual does not satisfy this condition then we cannot represent his evaluation of uncertain prospects in the way necessary for Harsanyi's theorem. While such a condition seems reasonable for individual's preferences it is less appealing when applied to social welfare. To see this, choose α as a distribution giving zero utility to one individual ($U_1 = 0$) and a utility level of one to another ($U_2 = 1$). Define the alternative β as a distribution giving $U_1 = 1$ and $U_2 = 0$. If individuals 1 and 2 are otherwise identical then it seems reasonable that social welfare should be indifferent between α or β; social welfare favouring neither 1 nor 2 over the other. Since we know the condition described above must hold for all δ, then we can choose $\delta = \alpha$ and construct $\bar{\alpha}$ which is α with probability 0.5 and α otherwise and construct $\bar{\beta}$ which is β with probability 0.5 and α otherwise. Harsanyi's analysis tells us social welfare must be indifferent between $\bar{\alpha}$ and $\bar{\beta}$. However this seems 'unreasonable'. Whatever the outcome in $\bar{\alpha}$ α must occur, which implies $U_1 = 0$ and $U_2 = 1$. In comparison $\bar{\beta}$ gives individual 1 a half chance of getting $U_1 = 0$ and a half chance of getting $U_1 = 1$. Similarly it gives 2 a half chance of $U_2 = 0$ and a half chance of $U_2 = 1$. $\bar{\alpha}$ favours 2 whereas $\bar{\beta}$ treats 1 and 2 identically. Thus, since Harsanyi's analysis implies $\bar{\alpha}$ must be indifferent to $\bar{\beta}$, in this example social welfare favours one individual over the other, which many argue is unacceptable (see Diamond, 1967).

In defence of Harsanyi and thus in defence of utilitarian social welfare one can argue that social welfare should only be concerned with the final outcome, and that in both $\bar{\alpha}$ and $\bar{\beta}$ someone will be rich and someone poor and that is all that should matter. The difference between the two can perhaps be best interpreted in a real world context by the following argument. If we know that income in society will be distributed unequally, does it matter whether this comes about through inheritance of wealth. Thus one can predict Lord X's son will be rich, even before his son is born, or through a random process that gives Mr and Mrs Smith's son an equal chance of being rich or poor. If one argues that the process of inequality is unimportant and that the inheritance process is no worse than a random process then one favours Harsanyi and hence utilitarianism. If one believes that the inheritance process is less fair than the random process then one rejects Harsanyi and hence rejects one of the strongest defences of the utilitarian concept of social welfare.

An alternative approach to the Harsanyi analysis would be to reject both the view that social welfare satisfied the 'rationality' conditions and the equal probabilities argument, i.e. believe choices should be made as if one had no knowledge of the probability of being rich or poor. This is basically the approach adopted by Rawls (1972), who argues that individuals in this position will not choose the utilitarian concept of social welfare. If one has to make a decision with no knowledge of the probabilities of the alternative outcomes then there are several standard decision rules which are advocated (e.g. see Chapter 24 of Baumol, 1961). One is to assume that each outcome has an equal chance of occurring (this would bring us back to Harsanyi's view). The difficulty with this can be seen clearly if one imagined a society with one rich person (call him 1) and ten poor people (numbers 2 to 11). If one defines the possible outcomes as being rich or poor using this rule one would assume an equal probability of wealth and poverty. If one took the same example but defined the possible outcomes as being individuals 1 to 11, then one has a probability of 1/11 of being rich and 10/11 of being poor. If one took the possible outcomes as being rich, being poor and tall, or poor and small, then the probability assigned to being rich would be 1/3 and that to being poor 2/3 (1/3 poor and tall plus 1/3 poor and small). Thus the assumed probabilities have no significance since they depend on how we choose to look at the pattern.

An alternative decision rule is to presume that whatever decisions one makes, the worst possible outcome will happen. If one believes this then actions will be taken which give the largest possible value to the worst outcome, i.e. one will maximise the welfare of the minimum outcome-maximin. The idea behind this rule is that given that one knows nothing about the possible probabilities at least one can set a limit to how badly off one can be when the state of the world is known. One could argue that individuals are more likely to worry about the worst outcomes when they are making important decisions which drastically affect their position. This is basically Rawls's argument, i.e. that when asked to choose a distribution acting as if one did not know one's position in society, because the choice is so fundamental and the outcomes so varied, individuals would place great emphasis on the worst off individuals to the extent that they would adopt a maximin strategy, i.e. maximise the worst off individual's welfare. In Rawls (1972) the arguments for maximin and other welfare considerations are given. Basically the principles Rawls believes individuals should adopt are the following:

1. Each person is to have an equal right to the most extensive total system of

equal basic liberty compatible with a similar system of liberty for all.
2. Social and economic inequalities are to be arranged so that they are . . . to
    the greatest benefit of the least advantaged, consistent with the just saving
    principle.

    Since we are considering distribution over time the saving principle is
the relevant piece of Rawlsian theory, but unfortunately the exact meaning of
this is not clear. The interpretation we shall adopt here is that savings should be
allocated to maximise the welfare of the worst-off generation (for alternative
interpretations see Dasgupta, 1974, and Grout, 1977a): thus the just saving
principle applies the maximum criteria to all generations. (For a general
discussion of the economic implications of maximin see Grout, 1977b). The
value of social welfare given any distribution is obviously the minimum level of
utility and thus social welfare is unchanged unless the worst-off individual
becomes better off. Rawls has suggested that the criteria should be extended by
looking at the welfare of the second worst-off individual and choosing the one
that makes him better off, if all the feasible distributions give the worst-off
individual equal utility. Similarly if all distributions give the two worst-off
individuals the same utility then Rawls argues that it is the welfare of the third
worst-off individual which decides the issue. For most purposes however it is
reasonable to focus purely on the maximin aspect and ignore these special
cases. Figure 6.8 shows indifference curves which depict distributions of U_1 and
U_2 giving equal social welfare using the maximin criterion. Point A and B have
identical social welfare since the worst-off individual has utility U_2^* in both

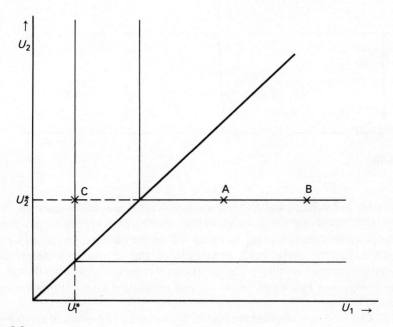

Fig. 6.8

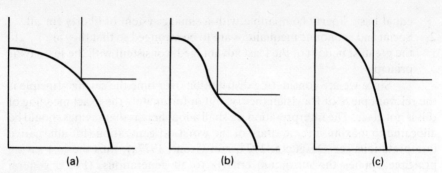

Fig. 6.9

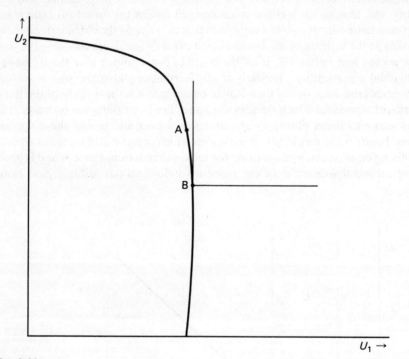

Fig. 6.10

cases. At C individual 2 has U_2^* but social welfare is now lower since one has $U_1^* < U_2^*$. Applying these social welfare indifference curves to our three examples shows clearly that equal utility will be the outcome in all cases (see Fig. 6.9). This property holds regardless of the productivity differences between generations, so that in Fig. 6.10 point B is the optimum even though A makes generation Two much better off and generation One not much worse off.

One wonders whether the people who advocate adoption of a maximin strategy with regard to the future would be quite so keen if we could return to the past and adopt the principle from pre-Industrial Revolution days, thus requiring us to make great sacrifices for the benefit of those past societies. Of course the past has made its sacrifices for our benefit and nothing can be done to change that. Thus if society adopted a maximin strategy we would benefit in the sense that we would only need to save enough to guarantee future generations are no worse off while we draw the benefit from the fact that past generations have saved greater quantities (in relative terms). Obviously it is only the first generation to adopt the strategy that can benefit in this extreme fashion and then only if they would otherwise have saved enough to make future generations better off than the present. It is when the present generation only need save a small quantity to maintain the present income per head over time (i.e. when the rate of return on investment is high) that the maximin concept of welfare is most criticised. In such cases the benefits to the present generation of a reduction in the savings rate is small in comparison to the reduction in welfare of the future generations. If, with an equal allocation of the exhaustible resource, other factors made for a very productive economy (say 10 per cent per annum) then 1 unit invested for the future (compounded annually) would generate over 2,000 units in eighty years' time. It is not unrealistic to think that a welfare function which, when income is equal for each generation, puts higher weight on £1 of our own consumption than on £2,000 worth of our great grandchildren's consumption, would be considered by many to be unfair. This is even more so when one considers that the welfare weight on our £1 would still be larger if the rate of interest were 50 per cent or 100 per cent. Once welfare levels are equal in each period there is no benefit that can accrue to future generations that can be large enough to offset the smallest possible loss today. This has led some people to suggest that maximin is unacceptable as a welfare function when analysing optimal distributions across time. The reason for this probably lies in the fact that there are an infinite number of generations, whereas at any period of time there are only a finite number of positions in society. The defence of maximin is given using examples of distributions across different individuals at one period of time. When we consider the set of generations then it seems implausible to concentrate social welfare on one generation (the worst off) and hence totally ignore an infinite number. Acting as if the worst outcome is most likely to occur is less reasonable if it is only one of an infinite number of possible alternatives all having a positive chance of occurring. Thus it is not unreasonable for someone to argue that one should maximise the welfare of the worst-off individual at any period of time but not maximise the welfare of the worst-off generation across time.

The arguments used by Harsanyi and Rawls have implicitly assumed that the comparison of welfare levels between two generations is well defined and costlessly determined. The opposite view would be to assume that it is impossible to compare welfare levels between generations and the only information one has concerning welfare levels is that a generation is better off or worse off given R_1 compared to R_2. If one believes this, then the elegant arguments out by Harsanyi and Rawls become operationally impossible and we must look elsewhere for feasible definitions of optimal distributions. Given this approach there is one view which has been discussed frequently in recent years, the concept of 'Fairness'. Loosely speaking, an allocation is called Fair if each generation or individual is better off with his own consumption than he would be with anyone else's: thus each individual feels he is better off than anyone else. More specifically a depletion path $(R_1 \ldots R_n)$ will be fair if it is Pareto-efficient and if $U_i(R_i) \geq U_i(R_j)$ for all i and j.

The simplest example of a Fair allocation has been known for thousands of years, and is the solution to the problem of how to divide a cake between two children so that each feels they have at least half of its value, i.e. neither child would swap his part of the cake for the other. If the cake is identical throughout then the answer is simply to cut the cake into equal sizes, but if the cake differs throughout, i.e. all currants in one area, etc., then in general equal sizes will not be Fair. In this case if one lets one child cut the cake and the other choose, the first child can cut the cake into two pieces each of which he thinks has equal value so that, whichever the other picks, the child who cut the cake must have half its value. The child who picks may not agree that the cake is cut into two equal value pieces but then he is free to pick the piece he values most, and so each child thinks he has at least half the value of the cake and they cannot benefit by swapping pieces. This is the simplest example of a whole series of problems which can be solved mathematically (see Dubins and Spanier 1961), the first part of which requires no mathematical knowledge). It can be shown that such distributions exist in exchange economies with many products for almost all preferences. Figure 6.11 shows a simple example in an Edgeworth box where x is obviously fair since both

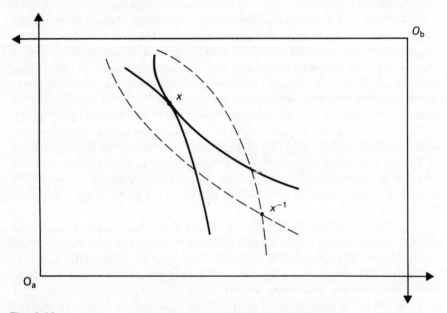

Fig. 6.11

individuals are better off with their allocation at x than if they swapped commodities (x^{-1}). More specifically it is intuitively obvious that if each individual's welfare can be represented as a function of one commodity and each individual always prefers more of this commodity to less, then the only fair allocation is the equal distribution of this commodity. Thus in the example we are looking at we can immediately state that the fair allocation must require an equal allocation of the resource at each period of time. The result is completely independent of the measure of individual welfare we use and thus avoids the hardest problem when comparing distributions across time. Operationally the procedure has much to recommend it and also has the property that whenever the utility function of all generations are identical the solution is identical to the maximin case, because in this example equal allocation of goods also implies equal allocation of welfare between generations. The advantage of the Fairness approach also brings with it the major disadvantage which is that even if one has some rough idea of a comparison of welfare levels the approach renders them irrelevant in the final allocation. The implications of this can be seen clearly in Fig. 6.12 where the allocation z is Fair but gives individual 1 almost as much of x as individual 2 and gives individual 1 almost all of y. If one could compare individual welfare then it may well be that 2 is much worse off than 1 and their welfare levels would be more equal at q. The Fair allocation is of course blind to this and there is no reason to prefer q to x. The equivalent problem arises in the example of two children and the cake. Here there may be an infinite number of ways of sharing the cake so that the children do not wish to swap but the Fairness criteria views all of these as equal. If it is clear that one

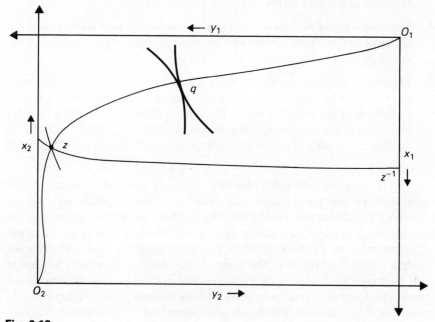

Fig. 6.12

child is much worse off than the other in utility terms no matter how the cake is distributed one may wish to criticise the Fair criterion for being indifferent between all the allocations, since one may feel the Fair allocation that gives most to the worse-off child is more 'fair' than all the others. If it is impossible to compare welfare levels then the Fairness criteria has much to recommend it. But if some comparisons are available one may feel it is pointless to ignore them.

The statement that the Fairness criteria is operationally easy to deal with does not imply that the maximin and utilitarian concepts are not. Indeed the main reason for concentrating on these three concepts is that they are all fairly easy to deal with, leading directly to weights which can be utilised in cost-benefit analysis. To summarise the basic implications of the approaches discussed:

1. Pareto efficiency – distributes the resource so that no one can be made better off without making someone else worse off. Still leaves the final choice to be made from an infinite set of possibilities, thus is best thought of as a necessary but not sufficient condition.

2. Compensation tests – bases welfare on hypothetical lump-sum transfers, thus totally favours the most productive generation and is not really acceptable to deal with exhaustible resource depletion, as it is discussed here.

3. Utilitarianism – maximising the sum of all generations' welfare, thus strongly favours those who are most productive in utilisation of the resource and utility terms.

4. Maximin – equal welfare levels for each generation, thus giving less of the resource to those generations who are most productive in utilisation of the resources and utility terms.

5. Fairness – equal consumption levels for each generation, thus giving less of the resource to those generations who are most productive in terms of utilisation of the resource but indifferent to differing utility functions. If one believes the welfare indicator for each generation is identical, then (5) is identical to (4) and can be interpreted as an alternative justification for (4).

To attain the optimal depletion path for any of the above statements of social welfare the government will have to influence prices and market structure. To add further justification for the necessity of this action and the necessity of a concept of social welfare I will close by discussing what the economic theoretic predictions are in a long-run resource market which is not 'guided' by the government. There are strong theoretical reasons to suggest that the long-run depletion of resources will not be Pareto efficient. Since Pareto efficiency is a relatively weak welfare statement most people would agree that if government action can make everyone better off then it is very acceptable. However, as we have seen, the concept is best interpreted as a necessary but not a sufficient condition. Thus the government would find themselves in a position of having to decide which of an infinite set of Pareto-efficient depletion paths they should 'guide' the economy to. This decision requires a concept of social welfare which brings us back to the necessity for such a concept and hence the necessity to discuss the alternatives.

The reason that the long-run depletion of resources may not be efficient rests on the presence of monopoly power. Monopoly power is of two types, the first being the monopoly power in the market for the resource, the theory of which is well discussed in any textbook. The second form of monopoly power is less discussed and more interesting in this context, and is the monopoly power which the present generation have over future generations. If the present generation wishes to consume all of the exhaustible resource in its lifetime then future generations, who are not born, can do nothing about it. The future generations themselves have power in the sense that if the present generation leave them quantities of resource in the belief that they will act in a certain fashion, the present generation cannot enforce this action and the future generations have freedom in how to distribute the resource they receive.

The model used in the paper as it stands does not throw much light on the problem since the present generation care only for their own consumption, and if left to maximise their 'selfish' welfare would use their monopoly position to deplete all of the resource in their lifetime. Such actions are of course unrealistic because the present generation do care about some future generations, e.g. children, grandchildren, perhaps great-grandchildren. It seems reasonable to assume that each generation will care to some extent about the consumption of their children and that there will also be generations in the future which they do not care about. To model this in the simplest form let us suppose that generation's utility function is now of the form

$$V_t = U(R_t) + U(R_{t+1})$$

thus caring for the next generation but no others. From the above discussion of monopoly power, generation t cannot influence what percentage of the resource left to generation $t+1$ is consumed by $t+1$, but of course generation t can totally determine the quantity passed on. Thus how much t will wish to pass on will depend upon the percentage $t+1$ wish to consume since any resource passed by $t+1$ to $t+2$ reduces the welfare of t. Taking the simplest form of the basic model (no economies of scale) and setting $n = 3$ we can determine the path that will be chosen. Let us start with generation Three who know that generation Four onwards receive $\bar{c}$ independent of the resource so that their best strategy is to consume all the resource they receive giving them utility

$$V_3 = U(R_3) + U(\bar{c})$$

The amount generation Two will wish to consume and the amount they will wish to pass on is now determinate since they know that generation Three will consume all they receive. Given the form of the utility function, generation Two will consume half of any resource they receive and pass half of it to generation Three. Thus we can now also solve the quantity generation One will consume since they will maximise their utility given that generation Two will consume a half of any resource they receive. The continuous line in Fig. 6.13(a) shows the depletion path associated with this, and in Fig. 6.13(b) the equivalent utility levels. However, it is easy to see that this path is not Pareto efficient. When generation One maximised their utility they will have chosen the amount of resource to pass on as the one that makes marginal utility of U_2 twice that of U_1 since generation Two only consume half of their resource. Suppose now the government entered and reduced the resource consumed at period 1 by a small amount and invested it so that it must all be consumed at period 2. Clearly this would make generation One better off since their loss will be the marginal utility of U_1 times the reduction at 1 and their gain will be twice the marginal utility of U_1 (i.e. the marginal utility of U_2) times the increase at 2. Generation Two must also be made better off since they are now consuming more of the resource at time 2 and consumption at time 3 is unchanged. Indeed generation

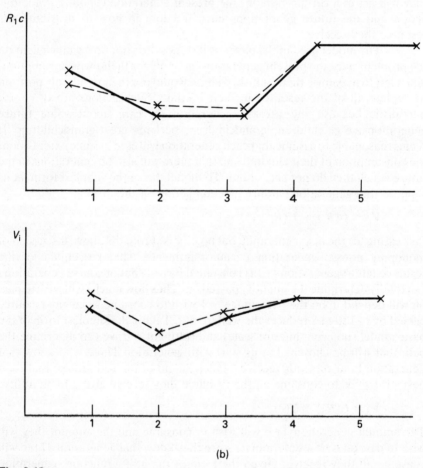

(b)

Fig. 6.13

Two will have become better off to the extent that they could fractionally reduce their consumption, passing this on to the third generation, and still have greater utility than in the initial allocation. This then allows generation Three to be better off as well. The consumption path and utility levels that result from this process are shown by the dotted line in Figs. 6.13(a) and 6.13(b). Thus government intervention to reduce the rate of depletion has made generations One to Three better off without making anyone worse off. The exact form of government intervention is obviously dependent on the concept of social welfare.

References

Baumol, W.J. (1952) *Welfare Economics and the Theory of the State*, Longman, London.

Baumol, W.J. (1961) *Economic Theory and Operations Analysis*, Prentice-Hall, Englewood Cliffs.

Dasgupta, P. (1974) 'On alternative criteria for justice between generations', *Journal of Public Economics* 3, pp. 405–23.

Diamond, P. (1967) 'Cardinal welfare, individualistic ethics and interpersonal comparisons of utility: a comment', *Journal of Political Economy* **75** pp. 765–6.

Dubins, L.E. and **Spanier, E.H.** (1961) 'How to cut a cake fairly', *American Mathematical Monthly* **68**, pp. 1–17.

Graaf, J. de V. (1957) *Theoretical Welfare Economics*, Cambridge University Press.

Grout, P. (1977a) 'Rawlsian justice and economic theory' in Jones, A. (ed.) *Economics and Equality*, Philip Allan, Oxford.

Grout, P. (1977b) 'A Rawlsian intertemporal consumption rule', *Review of Economic Studies* **44**, pp. 337–46.

Harsanyi, J. (1955) 'Cardinal welfare, individualistic ethics and interpersonal comparisons of utility', *Journal of Political Economy* **63**, pp 309–21.

Peacock, A.T. and **Rowley, C.K.** (1975) *Welfare Economics: A Liberal Restatement*, Martin Robertson, Oxford.

Rawls, J. (1972) *A Theory of Justice*, Oxford University Press.

Robbins, L. (1938) 'Interpersonal comparisons of utility: a comment', *Economic Journal* **48**, pp. 635–41.

Part Three ⸻

The fishermen will mourn and lament . . . and they will
languish who spread nets upon the water.
Isaiah 19:8
The economics of self-reproducible resources
management: the sea fishery

Introduction _____

The problems of managing an open-access resource have received increasing attention over recent years. Problems of environmental abuse can be seen partly as being the over-exploitation of an open-access resource. The other main thrust of economic work in this area has been into the preservation of non-appropriated stocks of animal species. Of these, the sea fishery has received most attention.

The development of the economic theory of fisheries exploitation took place in two phases. The comparative static model was initially developed by Gordon (1954) and remained the basis for analysis and policy recommendations alike. In the late 1960s, however, the application of the more searching tools of dynamic analysis to the problems of fisheries management led to the development of further insights into the problems.

The two authors whose papers are presented in this section are in the vanguard of these movements. Professor Copes concerns himself with the comparative static model of the fishery and develops from it a series of policy recommendations. But Copes goes further, and, from experience, draws together important considerations about the institutional problems related to implementing the policy measures that flow from the model. (It is often the failure of those responsible for regulation to grasp the relevance of a particular policy measure, rather than an inherent weakness in it, that leads to its not being adopted.)

Professor Munro develops the explicitly dynamic theory of sea fisheries. The introduction of this dimension leads to the moderation of several of the conclusions of the static model, particularly the relative positions of the population that yields the maximum biomass *vis-à-vis* the 'economically optimal' population level (depending on the rate of time preference for the society who is assuming responsibility for managing the stock). Munro is also concerned with adjustment problems.

It is this concern for adjustment problems and institutional constraints that distinguishes these papers. Economists are too often content simply to identify stable equilibria that are, in some sense, optimal, but do not concern themselves with 'how to get there' (stable equilibrium point) 'from here' (current disequilibrium situation). The attention that both authors give to such problems is commendable.

References

Gordon, H. Scott (1954) 'The economic theory of a common property resource: the fishery', *Journal of Political Economy* **62** pp. 124–42.

Chapter 7

Rational resource management and institutional constraints: the case of the fishery
Parzival Copes

The problem of the fishery

Throughout the world the fishery gives evidence of being a peculiarly troubled industry. While newly developing fishing operations sometimes show brief periods of extraordinary profitability, mature fisheries frequently are found in a depressed state. Economists have traced the problem of the fishing industry to the unique 'common property' characteristics of the fishery resource.[1]

Until quite recent times most fisheries have been exploited under conditions of unrestricted access. Typically, any person or group with the necessary gear has been allowed to dip into the common pool of fish. But what is everybody's resource in general, is nobody's property in particular. No one fisherman is personally motivated to conserve the resource, for any fish he would return to the water to grow to larger size will likely end up in the nets of a rival fisherman. Any expense he would undertake to conserve or enhance the fish stock, or to improve general fishery facilities, will yield him a negligible return. Most of the additions to the catch or improvements in returns that he would cause, will be enjoyed by other fishermen. Where no individual is able to recoup an investment made in the fish stock, everyone will personally incline to neglect the future of the resource. Each fisherman will be induced to take as much fish as quickly as he can, before others beat him to it. In this scramble for fish the stocks are often badly depleted and occasionally threatened with outright destruction.

Given the nature of the fishery problem, governments have often been called upon to regulate fishing activity. The aim has been to protect the resource from physical depletion and to moderate the economic damage of unbridled competition for shares of the common property resource. Where a fish stock spends its entire life within the boundaries of a single state, it is possible for that state to exercise unambiguous jurisdiction in regulating a fishery. However, the narrow three- or twelve-mile territorial limits that most countries claimed until recently, brought but a minor part of the world's marine fish stocks totally within the jurisdictions of single states. In international waters no authority has existed with effective power to institute and enforce fishery regulations. Voluntary agreements among states to regulate joint fisheries in international waters, on the whole, have not been very effective, because of unresolved rivalry among the participants and the constant threat of unrestricted fishing by nations that are not party to the agreement.

During the past two decades economic analysis applied to the problems of fisheries has brought forward new prescriptions for rational resource management. And, in a parallel development, the mounting pressures for extension of national fisheries jurisdictions in 1977 culminated in a rash of proclamations of 200-mile fishing limits by coastal states. It now appears that this limit has become an accepted standard of international law. As a result, the greater part of the world's commercial fish stocks have suddenly been brought under the legal jurisdiction of individual states. Together, these two developments – crystallisation of analytical understanding and establishment of effective legislative authority – justify the expectation that rapid progress may now be made in the formulation, application and enforcement of rational measures of fisheries management.

This article will attempt to outline the major requirements of optimal economic management and compare these with the experience of past efforts in fisheries regulation. Particular attention will be drawn to institutional constraints that in the past have obstructed rational management of the fishery resource. The analysis is designed to provide insights regarding the obstacles that will have to be cleared if new and more comprehensive attempts at fisheries management are to achieve a greater measure of success.

The economic theory of open access fisheries

The elementary economics of fisheries may be explained with the aid of Fig.7.1, which illustrates the basic long-run relationship between fishing effort and the catch taken in the exploitation of a single stock of fish under conditions of open access.[2] The relationship is expressed through the 'yield curve' (OAF), that shows the long-run (average) annual catch obtainable at different levels of fishing effort. The latter may be measured, for instance, in terms of 'vessel-years', where each unit represents the effect of a single vessel of a standard fishing capacity operating for a full season (year).

The sigmoid shape of the yield curve may be explained thus. At low levels of annual effort the catch naturally will be small in absolute terms. Increasing the level of fishing effort initially will bring larger catches. But the increase in the annual catch will be less than proportional to effort. There are three reasons that may account for this. The greater the number of vessels participating in the fishery, the more the fish stock will be thinned out, so that catches per vessel will decline. Also, as increasing effort removes more of the faster growing age groups, the rate of weight recovery of the stock will diminish. Finally, the reduction in the fish stock resulting from greater fishing effort will lead to diminished spawning activity, which may result in a decline in the number of juvenile fish growing to maturity.[3] At higher levels of fishing effort these effects may be so severe that in the long run the total size of the annual catch will actually decline, despite the greater effort expended. Thus, with increasing effort, the yield curve may be expected to rise at a declining rate until a 'maximum sustainable yield' is reached (AB). Beyond this point further

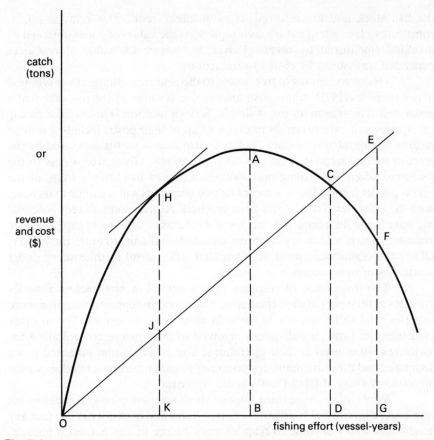

Fig. 7.1

increases in the level of annual effort will result in lower total annual catches.[4]

The yield curve of Fig. 7.1 may be turned into a revenue curve for the fishery concerned if a constant price prevails – simply by converting tons of catch to dollars of revenue at the prevailing price. A cost curve (OE) may also be drawn. Where the fishery is conducted by a fleet of similar vessels, with standard costs of operation and standard catch effectiveness, constant costs per unit of effort may be assumed. This will give total costs proportional to effort, resulting in a straight-line cost function passing through the origin. In accordance with the conventions of economic analysis, 'costs' here include all opportunity costs of the factors of production employed, including a 'normal' reward for labour and a 'normal' rate of return to capital invested.

In the situation portrayed in Fig. 7.1 the fishery will provide more than normal returns at effort levels between O and D. The net returns here are measured by the vertical distance between the revenue and cost curves. Thus at effort level OK there will be a net return of HJ, after accounting for all costs (JK) including the normal rewards that labour and capital could expect to receive in alternative uses. The net returns may be attributed to the qualities of the fish stock and are referred to as 'resource rent'. This rent, in effect, constitutes a free gift of nature, as it represents the values of goods obtained by mankind (measured by revenue) over and above the value of resources expended (measured by costs of production).

Under conditions of free access to the resource, fishing effort is bound to rise to the level OD, where costs are equal to revenue and no resource rent is generated. The reason for this is simple. So long as effort is below OD, fishing enterprises will earn revenues in excess of all of their posts, including normal returns to capital and labour. They will earn excess profits measured by the amount of resource rent that has been generated. Given free access to the resource, additional fishing units will be attracted to obtain a share of the excess profits (rent). The number of fishing units thus will continue to increase until the effort level of OD has been reached. At that point all rent potential will have been dissipated and only normal returns to labour and capital will be available. There will be no incentive for additional units to enter the fishery. OD is the equilibrium level of effort that will prevail in the fishery under conditions of open access.

The dissipation of resource rent potential in open access fisheries provides at least part of the explanation why economic opportunities in mature fisheries tend to be inferior to those in most other industries. For in other industries the firms or individuals involved usually own or control all of the major resources used in their operations. The available rent potential is not dissipated and there are many opportunities to earn returns over and above the opportunity costs of labour and capital employed.

While a fisherman must tolerate rivals who scoop away fish before his very eyes, a farmer holds title to his fenced fields and does not have to face any neighbours who will come to reap his grain before he can harvest it himself. The farmer may thus improve the returns from the resource he controls and keep the higher earnings for himself. The fisherman, however, is powerless to exert similar control over the resource he utilises. The problem in the fishery, it appears, is related to the absence of property rights exercised through 'sole ownership' (Scott, 1955).

There are additional reasons why mature fisheries often tend to be in a relatively depressed state. During the expansion phase of a developing fishery, immediate catches tend to be larger than later ones with the same level of effort. In terms of Fig.7.1, when effort first reaches OD, the catch will be in excess of CD, because the stock of fish available is of a size matching previously lower levels of effort. Only after several years of fishing at effort level OD will the fish stock be reduced to a size where it will yield no more than an annual catch of CD. In the meantime, therefore, revenues at OD will tend to exceed costs, so that additional fishing units may be attracted. Thus the fishing industry in its expansion is likely to overshoot the equilibrium effort level of OD. It may end up at effort level OG, where costs in the long run will exceed revenues by EF, so that less than normal returns will be available to labour and capital.

Below normal returns should induce operators to leave the industry, until the effort level has been brought back to the long-run equilibrium level of OD. However, entry to and exit from a fishery tend to be asymmetrical. It is usually not difficult to acquire a boat and equipment to join a fishery. However, once committed to the fishery, vessels represent sunk capital costs that usually cannot be recovered in full by withdrawal. Also, men who have established themselves in fishing communities often cannot withdraw, save at considerable expense. Labour and capital are frequently trapped in overexpanded fisheries where they earn less than normal returns. This depressing condition may be particularly severe where fishing communities are geographically isolated, with a few local alternative employment opportunities.

The criteria for optimal management

It is evident that open access to a common property fishery resource generally results in non-optimal conditions of exploitation. In terms of economic theory we may speak here of 'market failure' or 'misallocation' of the factors of production. From an economic standpoint, optimal conditions of exploitation would prevail at an effort level of OK where the net returns (rent) from the fishery would be maximised.[5] For at this level of effort the value of the catch (HK) exceeds the necessary costs of production (JK) by the largest possible amount (HJ).[6]

An increase in effort beyond OK is undesirable because for each additional unit of effort the additional revenue generated will be less than the additional costs incurred, so that a marginal loss will be sustained. It would therefore bring better returns to society to release any factors of production surplus to the needs of the OK effort level. For then they could be employed in alternative uses elsewhere in the economy, where they could earn their full opportunity costs. If effort is below OK, on the other hand, it will be socially desirable to expand effort to OK because the additional factors of production employed will increase revenue by more than they will increase costs. Economists therefore speak of the 'optimum sustainable yield' (OSY) at effort level OK, in contradistinction to the 'maximum sustainable yield' (MSY) at effort level OB. The MSY yields a maximum only in terms of the physical size of the catch, ignoring the costs necessary for its achievement. The OSY, in the other hand, is based on economic criteria. It indicates the 'maximum (net) economic yield' and may therefore also be referred to as the 'MEY'.

Given the fact that most fish stocks are public resources, limitation of effort to the optimum level usually can be achieved only by a public authority. This is a task that therefore commonly falls to the government. In limiting effort and determining who shall have access, the government essentially is exercising property rights on behalf of the public. In managing effort at the optimum level, the government may simulate the benefits of sole ownership of a resource in achieving maximum net economic returns (Copes, 1972).

The limitation of fishing effort

As indicated, economists have concluded that the major requirement for bringing about socially rational exploitation of the fishery resource is to limit effort to the optimal level where the resource rent may be maximised. In the process of such rationalisation two important questions arise:

1. How should the rents generated by effort limitation be distributed?

2. How can access to the fishery be controlled in a socially and politically acceptable way, without generating distortions that lead to non-optimal allocation of the factors of production?

The first of these two questions is primarily a matter of social value judgement and political decision. It may be contended that as the fish stocks are a public resource, any net returns in the form of rent should go into public revenue so that they may be used for the general benefit of society. The rents could be appropriated by the government by any one of a number of means. Rights to exploit the stocks could be auctioned off to the highest bidders. Licence fees could be set at levels so high as to absorb the rents. Royalties could be levied on the fish catch to a similar end.

It was observed above that fisheries are frequently found in a depressed state, characterised by low incomes. The social purpose of managing fishing effort under these circumstances often is precisely to secure higher incomes for fishermen. This argues for the allocation of some or all of the rents generated to the fishermen concerned. It may be achieved by restricting the fishery to a limited number of licensed fishermen so that rents will emerge which the fishermen will be allowed to retain. In the process, however, additional problems may arise in deciding who should have access to the limited number of fishing licences issued.

The emergence of conservation measures

A good understanding of the economics of fisheries exploitation has developed but slowly in industry and government circles over the last few decades. And this understanding is still far from complete. The phenomenon of stock depletion, however, has long been noticed and acknowledged. In this matter, the concern of both industry and government was aroused long ago. In the absence of analytical economic insight, the problem was seen simply as a natural phenomenon requiring scientific solution. Early measures of fisheries control, in consequence, tended to aim directly at conservation of fish stocks, with scant attention to the economic consequences of alternative measures of regulation. This is strikingly illustrated by the history of regulation in the Pacific salmon fisheries of the United States and Canada, which serves as 'a study of irrational conservation'.[7]

The life pattern of salmon makes them particularly vulnerable to stock depletion. As 'anadromous' fish, they spawn in fresh water but spend most of their life in salt water. The five North American species of Pacific salmon have precisely timed life cycles. Each 'race' of biologically interacting individuals of a species is spawned in a particular set of gravel beds of a river system. As juvenile fish they descend their native streams to carry out a wide-ranging feeding migration in the ocean. At full maturity (two to six years, depending on species and race) they reassemble at river mouths in preparation for spawning. The individuals of each race return as a group to the particular set of spawning beds where they originated. After spawning Pacific salmon invariably die.

The dense schools of salmon assembling in the estuarial waters and moving up the rivers together must in effect 'run the gauntlet' to reach their native streams. In these spawning concentrations the fish may be taken with particular ease. Indeed, salmon are among the few marine species of fish for which whole stocks may be readily exterminated by fishing effort. Traps, weirs and other catching devices may be so positioned as to barricade spawning rivers completely and capture entire races of fish. In the early days of the Pacific salmon fisheries (in the latter part of the nineteenth century) some stocks were thus entirely destroyed.

Among the first government regulations applied in the Pacific salmon fisheries was a prohibition on the installation of any barricades in salmon rivers. While this helped to ensure the escapement of fish for spawning purposes, it also curbed the use of the most efficient catching devices (traps) by banning them from their most advantageous locations. An economically effective requirement would have been to regulate the use of traps in rivers so as to ensure a level of escapement that would have maximised the next spawning run. But regulating the use of traps would have required authorising particular users – which, in turn, would have meant limiting entry to the fishery. This step was unacceptable to prevailing notions of free enterprise which were thought to imply unrestricted entry.

The banning of traps from river locations – and their subsequent complete prohibition – did not ensure adequate escapement of salmon. For the mobile gear of trollers, gill netters and purse seiners, operating in the open water approaches to salmon rivers, proved quite capable of overexploiting the stocks. Throughout the North American salmon fisheries additional regulations were introduced eventually to ensure more adequate escapement. Officials were given power to close designated areas to fishing and to suspend fishing activity to designated times. The rules could be applied at short notice. On scientific advice, area and time closures were thus manipulated to produce the desired escapement for the various races of fish on their spawning runs.

The regulations succeeded in maintaining more adequate escapement and thus helped to ensure a better long-term supply of fish. But area and time closures greatly reduced the efficiency of individual vessels by leaving them idle during much of the fishing season. It was common for the principal fishing areas to be open only for two or three days per week. With fixed costs spread over a smaller number of fishing days, the costs of operation per fishing day increased. Of course, more fish was available on the days open to fishing. But if this resulted in any improved net returns to fishermen, the benefits were quickly dissipated by the entry of additional vessels to the fishery.

The essential problem with the regulations was that they were based on a judgement of needs and a measurement of achievement by biological criteria only. The implicit test of success was whether the fishery came closer to producing the maximum sustainable yield of fish. From a simple biological perspective, failure to achieve the MSY meant 'wasting' fish that could be made available to mankind. There was much less comprehension that the use of excessive amounts of manpower and capital in the fishery meant 'wasting' other resources that could increase net benefits to society in alternative uses.

Gear restrictions

The biologically oriented regulations in the salmon fisheries did nothing to relieve economic pressures on individual fishing enterprises. With open access prevailing in the fishery, any emerging net benefits to fishermen were quickly eroded by additional entry. Under these circumstances the introduction of new technology that would divert fish to those operators first able to apply it, was seen as a serious threat to the interests of fishermen using conventional methods and gear. Moreover, any new equipment that would catch more fish would exacerbate the industry's aggregate pressure on fish stocks and erode the effect of existing regulations designed to ensure adequate escapement. For these reasons both the majority of fishermen and of officials in charge of regulations were bound to view technological innovation with distrust, if not hostility. In many fisheries around the world such circumstances led to the proscription of new technology.

The Pacific salmon fisheries offer several examples of regulatory action against the introduction of more efficient fishing techniques. One instance was in the gillnet fishery. The coarse-stranded nets commonly used in this fishery were avoided by salmon in daytime as they were quite visible in clear ocean waters in daylight. In consequence, gill-netters were confined to night-time fishing, except in some muddy estuarial waters with adverse fishing conditions owing to crowding and tidal action. When thin-stranded monofilament nets became available, that would catch fish in clear water in daytime as well as at night, they were quickly prohibited throughout the Pacific salmon fisheries of Canada and the United States.

With similar effect, a prohibition was issued against the use of electronic fish-finding gear. Such equipment would have been a boon to purse seiners. These vessels rely on the location of salmon schools in open waters where they may be encircled by ring nets. The fishing industry was also forbidden the use of aircraft or helicopters to assist in fish spotting. Fish traps eventually were banned everywhere in the Pacific salmon fisheries. And in Alaska entry to the fishery by vessels over fifty feet in length was prohibited. The most spectacular blow against efficiency, perhaps, was a regulation in the Alaskan fisheries – lasting until the 1950s – that barred gill-netters in Bristol Bay from using engines on their vessels. Thus, in comparatively modern times, the technologically most advanced nation in the world sent its fishermen out onto the Bering Sea in sailboats!

Fisheries rationalisation

Whether designed to achieve biological objectives of stock maintenance, or to protect different interest groups within the fishery from one another, the regulations of open-access fisheries have represented a strong institutional bias against technological efficiency. This bias has favoured labour-intensive distortions in the allocation of productive resources to fishing activity. Often this has been defended precisely by reference to the employment-generating effects of the regulations. This contemporary version of the Luddite argument has no more (or less) analytical validity than the old contention that the workers' interests require the destruction of machinery. With technological displacement of workers, there are of course problems of labour mobility that demand attention and solution. But, essentially, there would appear to be no more merit in retaining economically excessive numbers of workers in the fishing industry than in other modernising industries.

By criteria of economic efficiency the extent of excessive employment of manpower (and gear) in the Pacific salmon fisheries has been particularly severe. Without even moving to the more efficient equipment and techniques that have been banned, most sectors of this fishery could take the available catch with less than half of the manpower and gear currently in use. And in some cases less than one-fifth of the effort employed would be sufficient to take the full catch. If the most efficient technology were permitted in the industry, catching costs could be reduced to a modest fraction of the level that has prevailed.

What the foregoing implies is that the potential net benefits of the salmon fishery, in terms of rent, are very large. This is evident also when one considers the substantial size of the stocks, their high value as a favoured food fish and the extreme ease of capture in a 'gauntlet' fishery. The necessary costs of catching the fish need be but a modest fraction (probably no more than one-quarter) of their landed value. On this estimate, the Pacific salmon fisheries of the United States and Canada that now yield catches worth some $300 million annually, would render net benefits to society (through government revenue or excess profits to fishermen and companies) of $225 million per year.

Understanding of the phenomenon of resource rent dissipation in the fishery has become sufficiently widespread during the past decade to have generated an urge for corrective government action in many places. Several countries have now established limited entry regulations for fisheries in which overexploitation has been particularly evident. Most often this has taken the form of barring entry to any additional fishing units. This action may help to counter the further erosion of net benefits in fisheries that are still yielding some economic rent. In the case of most fully developed fisheries, however, open access has already had the effect of fully dissipating available rents. Introducing rules of entry limitation in such circumstances amounts to 'closing the barn after the horses are gone'.

To regenerate rents for fisheries in which previous open access has led to the build-up of excessive fishing capacity, it is necessary to reduce that capacity by removing fishing units from the industry. The first notable attempt in this direction was undertaken in 1968 in the Pacific salmon fishery of

Canada.[8] This was an appropriate setting for a major experiment, in view of the particularly spectacular rent losses resulting from open access in the salmon fishery. Not surprisingly, similar plans to remove excess capacity have recently been introduced for the salmon fisheries of the adjacent states of Washington and Alaska in the United States.

The salmon fishery rationalisation programme for Canada's west coast started with a ban on entry of additional fishing units and the withdrawal of licences from vessels that had not fished seriously for a few years. A further number of vessels that had only minor salmon catches were given a maximum period of ten years to withdraw from the fishery. The remaining vessels constituted the bulk of the salmon fleet, which had accounted for almost all of the catch. The owners of these vessels were considered to have established an entitlement to a share in the fishery and were given permanently renewable licences for their vessels. They were allowed to transfer these licences to replacement vessels or to sell their vessels with licences attached. Thus almost the full capacity of the salmon fleet was authorised, in the first instance, to continue with the fishery.

To induce a reduction in the capacity of the remaining salmon fleet, while maintaining equitable treatment of established fishermen, a programme was introduced to stimulate voluntary retirement of salmon vessels against compensation. Owners were given the opportunity to sell their vessels with licences attached at a price consisting of the appraised value of the vessel plus an incentive bonus. The programme was administered by an officially established 'Buy-Back Committee'. The vessels bought up in this fashion were auctioned off with the express condition that new owners could not obtain fishing licences for them.

The purchase of vessels by the Committee was funded by increased annual licence fees levied on salmon vessels remaining in the industry, plus the proceeds from the auction of vessels bought back. Increased licence fees were warranted as the removal of the bought back vessels increased average catches and incomes for the remaining fishermen. The buy-back programme thus was self-financing.

Increased average net returns for Canada's west coast salmon fishermen were proof that the buy-back programme had re-established some rent in the fishery. Further evidence of this lay in the fact that salmon fishermen were soon able to sell their vessels with licence attached to new aspiring fishermen at prices that were substantially in excess of the appraised values of the vessels. Thus the expectation of a stream of future rents to be earned by a salmon licence resulted in the 'capitalisation of these rents in the licenced vessel's sale price. The rent component of this price consisted of the discounted aggregate value of the stream of future rents.

Success in a buy-back programme tends to become self-exhausting. The modest incentive bonuses offered on vessel sales to the Buy-Back Committee in time were overtaken by increased licence values. Retiring salmon fishermen then found it more profitable to sell their vessels with licences attached to other persons who intended to continue with the fishery.

As a result the buy-back programme was suspended after three years of operation. Its continuation would have required raising the incentive bonuses substantially. In turn this would have necessitated an unpopular major increase in licence fees, or other revenues extracted from the industry, to keep the buy-back operation self-financing. With the large volume of potential rents that remained to be regenerated, it is evident that a determined continuation of the buy-back programme could have yielded substantial further increases in rents.

Controlling the quality of effort

Regulations to restrict entry have now been imposed in respect of some fisheries in a number of countries. In several instances the generation of resource rents is observable. In these fisheries a common experience is emerging. Wherever the number of fishing units is limited, their operators have attempted to build additional capacity into their individual fishing units. This has taken the form, among others, of increasing the size and speed of vessels, of adding more fishing gear or manpower, and of installing more sophisticated fish finding or navigational equipment.

Essentially, this is a manifestation of the same economic logic that attracts excessive numbers of vessels to an open-access fishery where rents are available. In both cases it signifies a competitive scramble for shares of the rent. Where the number of vessels is limited by regulation, the operators of individual vessels are driven to improve the catching power of their vessels in order to secure larger shares of the catch – and thus of the rent. Collectively, this proves a self-defeating effort, tending towards dissipation of the available rent. For the improvements in the catching capacity of individual vessels will increase the costs of operation of the fleet, while the stock of fish available remains unaltered.

Depending on technical progress and the cost-effectiveness of applying innovations to individual vessels, capacity improvement will proceed more or less rapidly towards full dissipation of the available rents. The question arises, naturally, as to what regulatory measures may be taken to arrest or reverse this rent dissipation. Two types of action appear available. The first is to prohibit directly the various forms of capacity improvement that are the cause of rent dissipation. This may often prove administratively difficult and prone to undesirable side effects.

To operate effectively fishermen need the power to manage their fishing units independently and to repair, improve and replace equipment where necessary or desirable. A tight regulation of all vessel alterations would be a nightmarish administrative task. It would also be destructive of the economic incentives and freedom of action that motivate individual fishermen to operate efficiently. In any case, vessel improvements stemming from technical innovation may have cost-reducing effects, such as diminished fuel consumption or reduced damage to gear. Such improvements should be encouraged in the interest of vessel operators, individually as well as collectively.

There is a need to distinguish between vessel changes that are cost-reducing and those that are capacity-increasing – allowing the former to proceed unhindered, while proscribing the latter. Unfortunately, many changes may be at once cost-reducing and capacity-increasing, for scale economies often accompany capacity increases. It may be difficult to decide in any particular case which aspect is the weightier. In any event, it is usually not possible to install effective regulations in anticipation of possibly undesirable innovations. Once they have been introduced to the industry it may be found inequitable to force their abandonment by operators who have incurred the expense of undertaking the innovation. And it might be considered equally inequitable to bar other operators from access to the same individually advantageous innovations.

The difficulties of managing technological change in a limited entry fishery may be largely circumvented by a continuing buy-back programme. This should allow direct regulation of change to be limited to the most obvious manifestations of undesirable capacity improvements – e.g. increases in vessel size without any benefit of scale economies. Fishermen could then be left free to experiment with any other improvements and innovations. Should they appear to involve some increase in total fishing capacity, this could be countered by the buy-back of enough fishing units to reduce total capacity to the desired level.

Rent allocation and intergenerational equity

Because mature fisheries so often are in a depressed state, characterised by low income levels, the introduction of measures of rationalisation frequently has as a major objective the improvement of fishermen's incomes. Leaving all or part of the rent generated in the hands of fishermen is a convenient way of achieving this objective. The generation of rent, as a rule, requires the restriction of effort, which is achieved by the issuance of a limited number of fishing licences only. If fishermen are allowed to sell their licences in fisheries that have generated rent, these licences will acquire a worth equal to the capitalised value of the stream of anticipated future rents. As observed above, this happened in the case of Canada's west coast salmon fisheries.

It is sometimes contended that the sale price of a licence represents a legitimate retirement bonus that will increase the lifetime earnings of a fisherman. However, the combination of a rent-generating limited entry regime and a permission to sell licences tends to defeat the purpose of increased lifetime earnings in the fishery on a long-term basis. For if licences are sold, they must also be bought. To meet the price of a licence, a new fisherman must borrow – or pay out of his savings – an amount equal to the capitalised value of the rents the licence may be expected to earn for him. When he does receive those rents later, they may have to be used entirely to pay off his debts or replace his depleted savings.

The privilege of selling limited entry licences may be expected to yield significant net benefits only to the first generation of fishermen to hold such licences in a particular fishery. Indeed, the first generation fisherman may be expected to enjoy a double benefit. He earns rents continually during his working life as a fisherman and then captures the value of all future rents to be earned by his licence when he sells it on retirement. Future generations of fishermen pay in advance for all of the rents their licences are expected to earn. Thus they enjoy no net benefits, unless during their working life additional unanticipated rents are generated, which are then in excess of the amounts needed to pay off the purchase price of their licences.

This inequity in the distribution of the rent between the initial and subsequent generations of fishermen is clearly in evidence in a number of limited-entry fisheries where significant rents have been generated. As a result, the subsequent generations tend to be excessively burdened by debt, so that the problem of inadequate net earnings in the fishery is regenerated despite the initial gains of rationalisation. The obvious way to avoid this problem is to prohibit the sale of licences. On retirement of one fisherman a replacement licence would then be issued free of charge to a new fisherman. Each generation of fishermen would then earn the rents accruing during their own working lifetime – no more and no less.

A ban on licence sales does generate another problem, however. If licences have an intrinsic value because of their rent-earning capacity, there are likely to be more candidates to take over licences than the number of replacement licences available. This requires the introduction of a system of licence allocation by the fisheries management authority. Often this is not a problem because there is a natural system of succession. Fishing operations are frequently conducted through small crews, consisting of master fishermen (owner/skippers), who hold fishing licences in their names, and deck hands. A seniority list of qualified deck hands, aspiring to become licensed fishermen, would provide for a ready system of licence succession.

The international dimension

Despite the advent of the 200-mile limit, with its establishment of national jurisdiction over important fishing areas around the world, serious international fisheries problems remain to be solved. Large areas of ocean continue to lie outside zones of national jurisdiction. In these 'high seas' there are a number of valuable fish stocks. The fishing pressure on these stocks is likely to increase, for the nations that have large 'distant water' fishing fleets, such as Japan and the USSR, find that they are being pushed out of many of the new 200–mile zones of coastal states that wish to reserve the resources of these zones for their own fishermen. Continuing utilisation of the displaced distant water fleets will require their increased employment in exploiting the remaining stocks of the high seas.

To deal with the many problems of international use of the oceans, the United Nations Organisation has called together the Third Law of the Sea Conference, which has been meeting in a continuing series of sessions since

1974. The Conference has produced a document to serve as a negotiating basis for an international agreement. The 'Informal Composite Negotiating Text' (ICNT) contains a set of articles regarding the use of the fishery resources of the oceans and proposes 'Exclusive Economic Zones' of 200 miles in which coastal states will exercise comprehensive jurisdiction over economic resources, including the fish stocks (United Nations, 1977).

While the work of the Law of the Sea Conference is being held up by a number of non-fisheries questions, a tabulation of national positions taken at the Conference indicates overwhelming support for the provisions of the fisheries section of the ICNT. The unilateral declarations of 200-mile zones by numerous nations signifies an anticipated confirmation of these zones in an agreement emerging from the Third Law of the Sea Conference.

The ICNT also addresses itself to the problem of management of the fishery resources of the remaining high seas and of transboundary stocks. These are stocks that straddle, or move back and forth across, the outer boundaries of national 200-mile zones or boundaries separating the zones of adjacent countries. The best the ICNT has been able to propose is that nations with common interests in these stocks come to an agreement on their use, conservation and management through bilateral treaties or multilateral fisheries conventions. But this is really no advance on the existing situation in international fisheries relations. Numerous fisheries conventions are already in existence (Koers, 1973). As in all matters of international law, they rely on voluntary agreement and compliance.

In international fisheries, the common-property-open-access problem remains unresolved. International law does not provide for any world authority that can prescribe and enforce rules of rational resource management and limit access to the resources of the high seas. The major difficulty lies in securing agreement to limit the build-up of excessive aggregate fishing capacity, as each country strives to gain a larger share of the international catch by increasing its own effort. And even when a number of countries agree among themselves on a formula for effort limitation, there is the constant threat of the 'free rider' problem. Other countries may join the fishery to capture an increasing share of the catch regardless of the limitations accepted by the signatories to the agreement.

Despite the poor prospects for rationalising resource use in the case of high seas fisheries, international fishery conventions are capable of yielding some benefits. If the competition for shares of the catch among participating nations prevents full optimisation in international fishery exploitation, it may still be possible to 'suboptimise' on agreed aspects of the fishery (Turvey, 1970). Thus countries may agree to use a particular minimum mesh size for their nets to allow smaller fish to escape and grow to maturity. For the same reason they may agree to close identified nursery areas for juvenile fish to all fishing. And to ensure an adequate spawning stock they may agree to a closed season for part of the year. All of these measures, if applied correctly, will tend to increase the total stock available for the participants in the fishery so that all may benefit together. Agreement on these measures may not be too difficult to obtain, particularly as it still leaves individual countries free to try for a larger share of the increased catch by expanding fishing effort. As such arrangements by their nature bar no countries from participating, the free rider problem is reduced, it is not eliminated, however. A few minor participants could profit from breaching the rules and taking fish from the protected sub-stocks at the expense of the majority – provided the majority were prepared to ignore the infraction and themselves continued to adhere to the rules.

The methods of suboptimisation suggested here would not prevent the resource rents in the fishery from being dissipated by excessive effort. However, with larger total catches some indirect benefits may be generated. Consumers could be expected to obtain larger amounts of fish, presumably, at lower prices. And on the production and marketing side, enterprises enjoying larger throughputs of fish products might benefit from scale economies.

The future of fishery resource management

Substantial advances in rational fisheries management may be expected in the years ahead. The 200-mile limit has now established a legal basis for management regulation over most of the world's commercial fish stocks, while the insights of economic analysis that have been developed recently have established criteria for rational management. Nevertheless, the unique common property characteristics of the fishery will continue to require unique adaptations to the institutions of a society in which sole ownership and control of resources is the norm.

There are three areas, in particular, where substantial further effort is required to overcome institutional constraints that impede rational exploitation of fishery resources:

1. Where fish stocks occur in areas of the high seas, or migrate across international boundaries, new forms of international accommodation need to be developed that will induce all countries affected to contribute to and adhere to a rational management regime in the common interest.

2. In limiting entry to and controlling effort in common-property fisheries, refined rules of management need to be developed so that necessary restraints on the level of effort will not interfere with initiatives in the industry to improve the quality of effort.

3. In mature fisheries, characterised by depressed socio-economic conditions, rationalisation of fishing operations, requiring reduced labour inputs, will have to be co-ordinated with strong efforts outside the fishing industry to provide alternative employment opportunities.

In fisheries management it is often very difficult to correct the errors of the past. It usually involves the need to reverse the effects of overexploitation. Stocks have to be rebuilt, vessels and gear have to be retired and men have to be accommodated in employment outside the fishery. But the knowledge that is now available about our irrational use of fishery resources in the past should help us to avoid repeating old mistakes in the future.

Notes

1. The first comprehensive analysis of the problem was presented in 1954 by Gordon.
2. The presentation here offers a simple 'steady state' analysis. The theory has been extended to a dynamic analysis, involving time discounting, by Clark and Munro. (1975).
3. In some 'non-self-regulating' stocks there is no perceptible connection between 'recruitment' (the access of juveniles to the stock of fish of catchable size) and the size of the parent stock. Because of high fecundity in these stocks there is always a great excess of juvenile fish and the number reaching catchable size depends primarily on other natural factors, such as food supply.
4. Note that in the short run catches will always rise with increased effort (additional effort in the same year can only result in an addition to the catch). The yield curve in Fig.7.1 shows long-run catches only. At each level this is the annual catch that will result from a sustained effort at that level for enough years to achieve a stable condition of the fish stock to match the level of effort. The higher the sustained effort, the smaller the resulting equilibrium stock in the long run.
5. This article abstracts from net economic benefits that may be obtained from the use of a fishery resource, other than resource rents. In an extended analysis Copes (1972) has shown how a combination of net benefits, including resource rent, consumers' surplus and producers' surplus, may be maximised.
6. The position of maximum rent may be determined geometrically by taking the point (H) at which a line parallel to the cost line is tangent to the revenue curve.
7. This phrase is the subtitle of a book by Crutchfield and Pontecorvo (1969), which treats in greater detail many of the regulatory practices of the Pacific salmon fisheries used as examples in this article.
8. For a more extensive account and analysis of this experiment see Pearse (1972).

References

Clark, C.W. and Munro, G.R. (1975) 'The economics of fishing and modern capital theory: a simplified approach', *Journal of Environmental Economics and Management* 2, 92–106.

Copes, P. (1972) 'Factor rents, sole ownership, and the optimum level of fisheries exploitation', *Manchester School of Social and Economic Studies* 40, 145–63.

Crutchfield, J.A. and Pontecorvo, G. (1969) *The Pacific Salmon Fisheries: A study of irrational conservation*, Johns Hopkins Press, Baltimore.

Gordon, H. Scott (1954) 'The economic theory of a common property resource: the fishery', *Journal of Political Economy* 62, 124–42.

Koers, A.W. (1973) *International Regulation of Marine Fisheries*, Fishing News, London.

Pearse, Peter H. (1972) *Rationalization of Canada's west coast salmon fishery: An economic evaluation. Economic aspects of fish production*,Paris: OECD, pp. 172–202.

Scott, A.D. (1955) 'The fishery: the objectives of sole ownership', *Journal of Political Economy* 63, 116–24.

Turvey, R. (1970) 'Optimization and suboptimization in fishery regulation', *American Economic Review* 54, pp. 64–76.

United Nations (1977) *Third Conference on the Law of the Sea*, Informal Composite Negotiating Text, UN Doc. A/CONF.62/WP.10.

Chapter 8

The economics of fishing: an introduction   *Gordon R. Munro*

As is true of all natural resources, fishery resources constitute capital assets from the point of view of society. Similar to man-made capital assets, such as factories and machinery, fishery resources are capable of producing a stream of returns to society through time. The prime difference between these resources and man-made capital, of course, is that the former come to us as gifts from nature.

While fisheries have the potential of generating a stream of returns to society through time, many of the world's commercially exploitable fisheries fail to do so. It has been realised by economists for almost twenty-five years that the chief reason for this apparently deplorable state of affairs is the fact that these fisheries are common property resources to which fishermen generally have easy access (Gordon, 1954). The fisheries are open to all who wish to exploit them and are owned by none. As such these resources stand in contrast to natural resources such as tracts of agricultural land which are subject to single or sole ownership.[1]

In this chapter we shall commence by asking what economic theory can tell us about the optimal management of a single fishery, when the resource is subject to sole ownership. We shall suppose that society owns the resource, that it is able to make its control over the resource effective, and that the resource is managed on its behalf by an all-powerful social manager. Having done this, we shall then go on to see why such a resource will be subject to mismanagement from society's point of view if, instead of being subject to sole ownership, it is treated as a common-property resource. Finally, we will outline some of the ways in which governmental authorities can attempt to alleviate the problems created by the common-property nature of the resource.

The reader is warned that these topics, if explored fully, can prove to be rather complex. Consequently, in this chapter we shall attempt to do no more than provide an introduction to them.[2]

We commence then by assuming that we are considering a fishery over which society is able to exercise effective property rights and which is managed on society's behalf by a social manager.

We have already stated that a fishery resource is capable of generating a stream of returns to society through time. We can now go further than this and say that a fishery is capable of generating a *sustainable* stream of returns through time. This arises because of the fact that a fishery, like a forest, but unlike minerals or fossil fuel, is a renewable resource. That is to say the resource is capable of growth. If for any given resource size, the harvest per period of time is just equal to the 'natural' growth, then the harvest or yield can be said to be sustainable.

To consider the question of growth more fully, we turn to one of the simplest of the marine biologists' fishery models, one usually associated with the name of Milner B. Schaefer (1957). This model is widely employed, implicitly if not explicitly, in the fisheries economics literature.

A fishery resource population, or biomass,[3] will grow as a result of new fish entering the fishery (recruitment) and as a result of growth of individual fish. The growth will be kept in check by natural mortality, e.g. natural predators. In the Schaefer model no attempt is made to distinguish among the aforementioned factors influencing the natural growth rate. Rather the factors are 'lumped' together and we talk of the growth of the biomass as being a function of itself and of the aquatic environment.

The aquatic environment places limits on the growth of the biomass by virtue of the fact that any given aquatic area will have a finite amount of nutrients. There will thus be an upper limit to the size of a fishery resource that a given area can maintain. We shall refer to this as the carrying capacity of the area.

We normally treat the aquatic environment as a constant so that we can express the growth of the resource very simply as

$$\frac{dx}{dt} = F(x) \qquad [1]$$

where x denotes the biomass expressed in terms of weight, t denotes time, and $F(x)$ is a given function representing the 'natural' growth rate of the biomass. Specifically in the Schaefer model we have:

$$\frac{dx}{dt} = r\left(x - \frac{x^2}{K}\right) \qquad [2]$$

where K denotes the carrying capacity and r, a constant, denotes the so-called *intrinsic* growth rate.[4]

We can graph the results as follows:

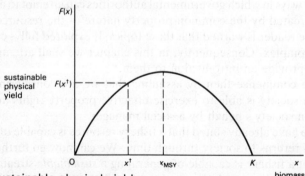

Fig. 8.1 Sustainable physical yield.

At very low levels of population, nutrients are abundant in relation to the fish population, thus the growth of the biomass, $F(x)$, initially increases with the population. However, as the pressure upon the aquatic environment increases, $F(x)$ reaches a maximum and then declines. It reaches zero at K – the maximum population size the environment can support. Here recruitment and the growth of fish are balanced off by natural mortality. The population level K is often referred to as the natural equilibrium population level.

Now let us introduce a new type of predation in the form of fishing or harvesting. Equation [1] must now be modified as follows:

$$\frac{dx}{dt} = F(x) - h(t) \tag{3}$$

where $h(t)$ denotes the harvest rate. If for a given biomass level, say $x^\dagger$, $h(t) = F(x^\dagger)$ and hence $dx/dt = 0$, we say that harvesting is taking place on a sustained basis. Hence $F(x^\dagger)$ can be viewed as the *sustainable* physical yield with respect to biomass level, $x^\dagger$.

Next let us consider the factors determining the harvest rate. The harvest rate will depend upon the size of the biomass, x, and the amount of labour and capital (fishermen and vessels) which are directed to the task of extracting the fish from the sea. It is common in fisheries economics to combine the labour and capital (i.e. assume fixed factor proportions) and refer to the package as fishing *effort*.
Thus we have

$$h(t) = h(E, x) \tag{4}$$

where E denotes fishing effort. A commonly used specific form of the above harvest production function is:

$$h(t) = q\, E^\alpha\, x^\beta \tag{5}$$

where q is a constant often referred to as the catchability coefficient. The exponents α and β are also constants.[5]

In the Schaefer model it is assumed $\alpha = \beta = 1$ so that we can re-write Equation [5] as

$$h(t) = q\, E\, x \tag{5a}$$

This form of the harvest production function is used extensively in the fisheries economics literature.[6]

Up to this point we have talked solely in terms of physical yields from the fishery. As economists, however, we are really interested in the net economic benefit the fishery is capable of generating for society over time. This means that we must look at the gross benefit to society of harvested fish minus the costs of harvesting.

Let us begin by introducing the following simplifying assumptions. The first is that the price of harvested fish in the market provides a good representation of the gross marginal social benefit of the fish to society. The second is that the demand for fish is perfectly elastic. Next we assume that the supply of effort inputs is infinitely elastic and finally that the unit cost of effort provides a good measure of the marginal social cost of fishing effort.

If we return to Fig.8.1, it can now be seen that the sustainable yield curve can be transformed into a sustainable revenue curve simply by multiplying $F(x)$ by the price of fish, p. By assumption the total cost of fishing effort is

$$C(E) = a\,E \qquad [6]$$

where a is the cost of fishing effort. If we assume that the relevant harvest production function is that given to us by Equation $[5a]$, then total harvesting cost is given to us by:

$$C(h, x) = \frac{ah}{qx} \qquad ^7 \qquad [7]$$

and unit (average) harvesting cost by:

$$c(x) = \frac{a}{qx} \qquad [8]$$

If we continue with the Schaefer model and hence assume that the growth function and harvest production function are those given by Equation $[2]$ and Equation $[5a]$ respectively, then it can easily be shown that the cost of harvesting the sustainable yield can be represented graphically as follows:[8]

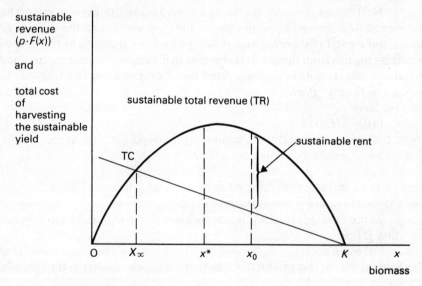

Fig. 8.2   Sustainable rent.

Note that in this model harvesting (output) costs are a function, not only of the harvest rate, but of the biomass level as well. It can be seen from Equations $[7]$ and $[8]$ that harvesting costs are an inverse function of x and indeed that harvesting costs become infinitely high as x approaches zero. One intuitive explanation for this relationship is that, as the stock or biomass diminishes, more time and effort must be spent on just searching for the fish. One common complaint from fishermen facing stocks that have been subject to heavy exploitation is that of declining catch per unit of effort (CPUE) with resultant increases in their costs per unit of fish harvested. Recognising that catch and harvest are synonymous, we can see from Equation $[5a]$ that CPUE is just qx.

Figure 8.2 shows the net economic benefit TR − TC that can be enjoyed from the fishery at different biomass levels on a *sustained* basis. The difference between TR and TC is commonly referred to as the sustained 'rent' generated by the resource. At the biomass level, x_0, where the slope of TC is equal to the slope of the sustainable revenue curve, sustainable rent is maximised. It is tempting to believe that maximisation of sustained rent is optimal from society's point of view. In fact this will virtually never be the case.

If sustained rent maximisation is not the optimal management strategy, what is? To find the answer we must recall the fact that a fishery is a capital asset and draw upon that body of economic theory known as the theory of capital.

The basic problem in capital theory is to determine the optimal stock of capital as a function of time, be it for the firm or, as in our case, for society.

A stock of capital may be adjusted by a process of investment, positive or negative. With respect to the fishery resource, investment can be said to occur whenever harvesting is carried on in other than a sustained yield basis. Return to Equation *[3]*. If $h(t) < F(x)$, dx/dt will be positive and the biomass will increase. By harvesting at less than sustained yield society is 'investing' in the resource. Similarly if $h(t) > F(x)$, the biomass will decline. Disinvestment, or negative investment, can be said to have occurred.

If we can assume that the fishery is small enough so that decisions to invest or disinvest in the resource will have no impact upon the national rate of net saving, then we can say unequivocally that a decision to invest in, to conserve, a particular fishery will come at the expense of other investment opportunities open to society. Thus in deciding to build up or run down a particular fishery resource, we must compare the marginal yield on the resource to society with marginal yields on alternative investment opportunities in the economy. We can state that we shall have achieved asset or stock equilibrium, i.e. we shall have achieved the optimal stock or biomass level, when the marginal yield to society on the fishery biomass is equal to the marginal yield on other capital assets of the same risk class.

We shall make the highly simplifying assumption of perfect foresight. Hence we can treat all real assets in the economy, including fishery resources, as essentially riskless. The common marginal yield or interest rate we shall refer to as the social rate of discount. Let us denote the social rate of discount by δ.

How do we express the marginal yield on the resource? The yield will in the first instance be related to the marginal change in sustainable rent brought about by an incremental change in the biomass level, x. If we increase x we will be rewarded by a stream of returns through time in the form of additional rent. For biomass levels below x_0 in Fig.8.2, $d\{SR\}/dx$ (where SR denotes sustainable rent) will be positive, i.e. there will be a positive reward for investment. At biomass levels greater than x_0, however, $d\{SR\}/dx < 0$. At $x = x_0$, of course, $d\{SR\}/dx_0 = 0$.

Let us refer to $d\{SR\}/dx$ as the marginal rent product of the biomass. To express this marginal return as a yield or interest rate, we divide $d\{SR\}/dx$, by the cost or 'price' of the incremental addition to the biomass or resource capital. The cost to society of making this 'investment' to the biomass is the marginal current net benefit forgone by not harvesting the incremental addition to the biomass. This in turn is simply equal to $p - c(x)$, the value of the incremental fish on the market minus the cost of harvesting the fish.[9]

Hence the marginal yield on the resource is:

$$\frac{d\{SR\}/dx}{p - c(x)} \qquad\qquad [9]$$

which can be referred to as the 'Own Rate of Interest' of the biomass. At any point in time, the optimal biomass level will be that level at which the Own Rate of Interest of the biomass is equal to the social discount rate. This we can express more formally as:

$$\frac{d\{SR\}/dx^*}{p - c(x^*)} = \delta \qquad\qquad [10]^{10}$$

where x^* denotes the optimal biomass level.

Equation [10] we shall hereafter refer to as the *Golden Rule of Resource Conservation*.

In Fig. 8.2 we show x^* as lying between x_0, the biomass level at which sustainable rent is maximised and x_∞, the biomass level at which sustainable rent is equal to zero. Why x^* lies within this range will become apparent below.[11]

What can we say about the optimal level of effort, E^*? The optimal level of fishing effort will be the *minimum* level required to harvest $F(x^*)$ on a sustained basis. Return to Equation [5a]. Upon substituting $F(x^*)$ for h and rearranging terms we have:

$$E^* = \frac{F(x^*)}{qx^*} \qquad\qquad [11]$$

We can now address ourselves to two questions raised earlier. First, how much validity is there in the claim that it will virtually never be optimal to maximise sustainable rent. Second, why will net economic benefit from the fishery or rent tend to be dissipated if society permits the fishery to operate on open-access, common-property basis.

With respect to the first question note that maximising sustainable rent implies that we have reached a stock level at which $d\{SR\}/dx - 0$. But this also means that the Own Rate of Interest is zero. Equation [10], the Golden Rule of Resource Conservation, would then be satisfied only if $\delta = 0$. Consequently, it will be worth society's while to maximise sustainable rent only if the marginal yields on all alternative capital assets equal zero. This possibility is so remote as not to merit serious attention.

With respect to perhaps the more important question of an open-access, common-property fishery, we first observe that a rational fisherman will, in contrast to society as a whole, have no incentive to invest in or conserve the resource. We have seen that the sole owner of a fishery can expect a positive marginal reward for investing in the resource whenever $x < x_0$ (see Fig. 8.2), in the sense that $d\{SR\}/dx > 0$. The individual fisherman in a common property fishery, on the other hand, must assume that whatever fish he refrains from harvesting will be captured by his competitors. His own perception of $d\{SR\} / dx$, therefore, is that it will be equal to zero.

The consequences of this perception can be seen clearly if we return to Equation *[10]*, but rearrange the terms so that it now reads as follows:

$$\frac{d\{SR\}/dx^*}{\delta} = p - c(x^*) \qquad [12]$$

The R.H.S. of Equation *[2]*, as we have already seen, is the price, or to be more specific, the supply price of an increment of 'capital' or biomass. The L.H.S. is a measure of what that increment of 'capital' is worth to the investor. It is the present value of the stream of additional sustainable rent made possible by the investment. We shall call this the demand price of an increment of capital. In equilibrium the demand price and the supply price will be equal.

Thus, saying that the rational individual fisherman perceives $d\{SR\}/dx$ as being equal to zero means that he perceives the demand price of 'capital' also to be equal to zero. He will therefore have an incentive to expand his fishing activities so long as $p > c(x)$, so long as the marginal revenue from harvesting exceeds the perceived cost at the margin of harvesting the fish.[12]

Now let us suppose we commence with a common property fishery in which, in fact, $p > c(x)$. Not only will existing fishermen attempt to increase their fishing effort, but new fishermen will be attracted into the fishery as well. For the rent which is being generated by the resource will show up as supernormal returns to both physical capital and labour.

The expansion of harvesting activity must invariably cause the global harvest to exceed sustainable yield. Hence the biomass level will fall. One fisherman alone cannot have much impact upon the biomass level, but all fishermen together will. As the biomass level falls harvesting costs will increase. The rising costs will appear to each 'firm' or fisherman as an external diseconomy. As the biomass continues to decline, $c(x)$ will steadily increase until $p = c(x)$. There will then be no further incentive for existing fishermen to expand their effort or for new fishermen to enter. The common property fishery will then be in equilibrium. We refer to this equilibrium as *bionomic equilibrium*.

In Fig. 8.2 bionomic equilibrium is represented by x_x. In our model the equality of p and $c(x)$ implies that total sustained revenue will equal the total cost of harvesting the sustained yield. Hence equilibrium in an open-access, common-property fishery implies that *sustainable* rent will be totally dissipated.

Would bionomic equilibrium ever be optimal from the point of view of a sole owner? We can answer this question by returning to Equation *[12]*. Bionomic equilibrium will be optimal *if* $\delta = \infty$, in other words if the sole owner completely ignores the future. If the sole owner (society) does not take this extreme position, then bionomic equilibrium in our model will always be non-optimal as it will imply an excessive disinvestment of the resource asset.

It is worth stressing that undesirable results of a common property fishery do not arise because of the stupidity of individual fishermen or because of their blind disregard for the future. The results arise because of a market failure, in the sense that the market sends out the wrong signals.

We can now see that both bionomic equilibrium, x_∞, and the biomass level at which sustainable rent is maximised, x_0, represent polar extremes and that each is unsuitable as a policy objective. The biomass level x_∞ would be optimal for society only if $\delta = \infty$; while x_0 would be optimal only if $\delta = 0$.

Let us return to the problem of common property fisheries and ask what in theory, at least, the governmental authorities can do to correct the situation, assuming that nationalisation of the fishing industry is politically unacceptable. The most obvious step the authorities could take would be that of restricting harvests. Let us return to Fig. 8.2, and suppose that when the social authorities take control of the fishery that it is in bionomic equilibrium at $x = x_\infty$. By various decrees, the authorities restrict the harvest to less than $F(x_\infty)$. The biomass is permitted to increase until it grows to x^*. From that point on the authorities effectively set the global harvest rate at $F(x^*)$. Then surely all will be well.

All will not be well, however, if there is an absence of restrictions on fishing effort. The optimal biomass level, x^*, was determined on the implicit assumption that the fish are harvested at minimum cost. If at x^* rents are generated, existing fishermen will struggle to expand their fishing effort in order to increase their share of the valuable catch. New fishermen will be attracted by the high returns. The authorities have control over the global harvest, but they do not have control over the harvest of any one vessel.

As a consequence far more vessels and fishermen than are needed to harvest $F(x^*)$ on a sustained basis will be engaged in the fishery. The TC curve in Fig. 8.2 will appear to rotate clockwise until it intersects the sustained revenue curve at x^*. In other words a new bionomic equilibrium will become established at x^*. This scenario is not far-fetched. One can cite many cases in which governmental authorities have restricted harvests, while failing to restrict fishing effort, and have then subsequently found themselves confronted with a grossly overcapitalised fishery.

What other measures can the authorities undertake? If, upon restricting harvests, the authorities find that fishing effort becomes excessive because of false market signals, then it would seem obvious that the authorities should prevent the fishermen from responding to these signals by restricting the amount of fishing effort. The authorities might place limits on the number of vessels which can engage in the fishery, or better yet, on the total vessel tonnage. Under the latter restriction new vessels could enter the fishery only if an equivalent tonnage of old vessels was removed.

All of this is easier said than done, however. For the incentive for individual fishermen to expand fishing effort and thus to find ways around the restrictions remains. Let it be supposed that vessel tonnage is restricted. How does one prevent the fishermen replacing old vessels with new vessels that have greater catching power? There is a great risk that most of the initial benefits offered by the restrictions will leak away over time.

Another approach that the authorities might take is to attempt to alter the market signals by taxing away the rent being garnered by the fishermen. For example, if one returns to Equation *[12]* and is able to calculate $\dfrac{d\{SR\}/dx^*}{\delta}$ using minimum harvesting costs, then one has the basis of an incremental tax on harvesting. Impose a tax T per increment of fish caught such that

$$T = \frac{d\{SR\}/dx^*}{\delta}$$

[13]

The fleet and the biomass will then be in equilibrium only when the following conditions are satisfied:

$$\left.\begin{array}{l} x = x^* \\ \text{and} \\ E = E^* \end{array}\right\}$$

[14]

When these conditions are satisfied the perceived rent at the margin for each fisherman will be:

$$p - [c(x^*) + T] = 0$$

[15]

If $x < x^*$ and/or if $E > E^*$ vessel owners will fail to cover their opportunity costs and the fleet size will diminish.

Yet another approach is to meet the common-property problem head on by assigning property rights to fishermen in the form of harvesting quota rights. Upon achieving x^*, authorities assign shares of the total allowable catch, $F(x^*)$, to individual fishermen. Now individual fishermen can increase their share of the catch only be purchasing the quotas of other fishermen. They will thus have no incentive to expand effort in order to increase their share. Indeed they will have an incentive either to harvest their share at least cost or to sell off their quota rights to their more efficient competitors.

All of these approaches have difficulties associated with them, technical or otherwise. A tax on catch, for example, would certainly arouse intense political resistence. Fishermen's quotas, on the other hand, create serious equity problems. Why should a few select fishermen enjoy all the benefit of effective fisheries management?[13] Prescriptions for effective fisheries regulations are easy to state, but extremely difficult to implement in practice.

There yet remains one important question which we have not raised. Let us assume that the authorities are somehow able to exercise effective management with respect to a fishery. Assume further that, when the authorities come to impose their management regime, the fishery is in bionomic equilibrium. How rapidly should the authorities move to the desired goal of x^*? Should they move there as quickly as possible by reducing fishing activity to the minimum, perhaps to zero, until the optimal stock level has been reached, or should they proceed more slowly?

The answer, not surprisingly, depends on the underlying conditions. If it is easy to shift men and vessels out of the industry and if the demand for harvested fish and the supply of effort inputs are both infinitely elastic, then we should indeed move as quickly as possible to x^*. In this situation we have nothing to lose by investing rapidly in the resource, rather than slowly.

On the other hand, there may well be situations in which the rapid approach to x^* is non-optimal. Let it be supposed, as is true in many fisheries, that the vessels employed have few if any alternative uses and that relocating fishermen in other parts of the economy is expected to be a slow and difficult process. Then it can be demonstrated that the optimal approach to x^* will be a slow and gradual one. Drastic reductions in harvests will be unwarranted.[14]

In this chapter we have attempted to provide no more than a brief introduction to the economics of fishing. As such we have made many simplifying assumptions. The assumptions of perfectly elastic demand and input supply functions are only the most obvious.[15] We have also assumed that the relevant parameters, e.g. the price of fish and the social rate of discount, are fixed through time. Obviously they are not. We have also assumed that the biological yields associated with given biomass levels are stable. They are not. Rather they must be expected to fluctuate and do so in a random manner.

These simplifications have been introduced deliberately, however, in order that the student can achieve a grasp of the fundamentals of fisheries economics. Once this has been accomplished, he or she can and should move on to more complex models in which the aforementioned assumptions are relaxed.[16]

Notes

1. The common property problem is not unique to fisheries, however. The problem appears in petroleum fields, fresh water resources and the environment.
2. For a further discussion of these topics, the student is urged to turn to Clark (1976) and Clark and Munro (1975).
3. The biomass or stock should be thought of as being measured in terms of weight rather than the number of fish.
4. The growth rate will approach r as x approaches zero.
5. The production function is recognisable as a Cobb-Douglas type production function.
6. We can, however, certainly deal with situations in which neither α nor β are equal to 1.
7. From Equation [5a] we have $h = q E x$. Hence:

$$E = \frac{h}{qx}$$

Thus substituting in to Equation [6] we have:

$$C(h, x) = \frac{ah}{qx}$$

8. From Equation [3] we have:

$$\frac{dx}{dt} = F(x) - h(t)$$

From Equations *[2]* and *[5a]* substitute in to Equation *[3]* for $F(x)$ and $h(t)$ giving

$$\frac{dx}{dt} = r(x - \frac{x^2}{K}) - q\,E\,x.$$

If the harvest is sustainable, then $dx/dt = 0$. Setting dx/dt equal to zero and solving for E we have:

$$E = \frac{r}{q}\left(1 - \frac{x}{K}\right)$$

We thus have an equation of the level of effort, E, required to harvest $F(x)$ on a sustained basis for any level of biomass x. Denoting the total cost of harvesting on a sustained basis as TC we have:

$$\text{TC} = a\,\frac{r}{q}\left(1 - \frac{x}{K}\right), \text{ where } a \text{ is the unit cost of } E.$$

9. The expression $c(x)$ is equal to $\dfrac{\partial\,C\,(h,\,x)}{\partial\,h}$

   Note that in our model:

   $$\frac{\partial\,C\,(h,\,x)}{\partial\,h} = \frac{C\,(h,\,x)}{h}$$

10. $\text{SR} = (p - c(x))\,F(x).$

    Carrying out the differentiation in Equation *[12]* we have:

    $$F'(x^*) - \frac{\partial\,C\,(h,\,x^*)/\partial\,x^*}{p - c(x^*)} = \delta \tag{12a}$$

    The first term on the L.H.S. of Equation *[12a]* can be viewed as the marginal physical product of the resource. The second term on the L.H.S. is referred to as the 'marginal stock effect'. It is a measure of the marginal benefit enjoyed from investing in the biomass by virtue of the fact that a larger biomass means lower harvesting costs (note that $\partial\,C(h,\,x)/\partial x < 0$.) See Clark and Munro, 1975, pp.95–6.

11. In the past biologists as managers used to focus on that stock level which would produce the maximum sustained yield, which we shall denote as x_{MSY}. Where does x^* lie in relation to x_{MSY}? The answer is that x^* can either be greater than, equal to, or less than x_{MSY}.

    If we return to the previous footnote we can see that, if both the marginal stock effect and the discount rate are equal to zero, Equation *[12a]* reduces to:

    $$F'(x^*) = 0 \tag{12b}$$

    Then x_{MSY} would be the optimal stock level. Introducing a positive marginal stock effect will push us towards stock levels greater than x_{MSY}. However, introducing a positive discount rate will push us in the opposite direction. Hence it is impossible to predict a *priori* which side of x_{MSY} x^* will lie.

12. See note 9.

13. One way of getting around the equity problem would be by having the government auction off the harvest quotas.

14. This is in fact a very difficult topic which has only recently been subject to serious analysis. See Clark, Clarke and Munro (1977).

15. If we drop these extreme elasticity assumptions, our discussion of the consequences of allowing a fishery to be exploited on an open-access, common-property basis must be modified substantially. See Copes (1972).

16. See note 2.

References

Clark, C.W. (1976) *Mathematical Bioeconomics*, Wiley, New York.

Clark, C.W., Clark, F.H. and **Munro, G.R.** (1977) 'The optimal exploitation of renewable resource stocks: problems of irreversible investment', *Resources Paper No. 8*.

Clark, C.W. and **Munro, G.R.** (1975) 'The economics of fishing and modern capital theory: a simplified approach', *Journal of Environmental Economics and Management* **2**, pp. 92–106. Department of Economics. University of British Columbia.

Copes, P. (1972) 'Factor rents, sole ownership, and the optimum level of fisheries exploitation', *Manchester School of Social and Economic Studies* **40**, pp. 145–163.

Gordon, H. Scott (1954) 'The economic theory of a common property resource: the fishery: *Journal of Political Economy* **62**, pp. 124–42.

Schaefer, M.B. (1957) 'Some consideration of population dynamics and economics in relation to the management of marine fisheries', *Journal of the Fisheries Research Board of Canada* **14**, pp. 669–81.

Part Four _____

But in this world nothing can be said to be certain, except death and taxes . *Benjamin Franklin*

Economic incentives and environmental management

Introduction

Benjamin Franklin's words concerning taxes as the other certainty of life highlight one reason why the use of economic instruments has come to supplement the regulatory approach to environmental management in many countries. Rarely is the suggestion made that, in practice, economic instruments can 'do the job' by themselves, 'the job' being to regulate the level of pollution to some predetermined level. From Turvey (1963) onwards the suggestion has been, rather, that a mixed regulatory/economic incentives approach may be optimal in a wide range of real world situations (Baumol and Oates, 1979). The economic aspects of a mixed policy confer on it two advantages:

1. It has efficiency built into it (i.e. a charge or tax imposed on polluters equal to the marginal social costs of their emissions at the desired level will be the least cost approach to achieving this desired level);
2. It is difficult to avoid or evade.

The two readings in this section address the application of the economic charge both to the solid waste problem and to air or water pollution. In the former case, Butlin discusses not simply the theoretical, allocative efficiency of user charges and product charges in a world with no economic friction, but rather some of the more pragmatic problems of actually implementing schemes that incorporate the use of these economic instruments. The purpose of his chapter is to show that, in practice, the theoretical gains claimed for economic incentives to aid in waste management may be realised, and not swallowed up in transactions and administrative costs. Marquand similarly shows that for a corrective Pigovian tax to be imposed in order to control water quality in a river basin, the required knowledge base is greater than the simple theoretical model suggests. She also raises doubts relating to the distributive effects of the tax and the relationship between pollutant concentrations in effluent and content in the water-body at large, two concerns that relate also to air pollution regulation problems.

If there is a moral to be drawn from these two chapters it is that from the viewpoint of economic efficiency, economic charges will improve environmental management. However, neither the administrative costs, nor the required knowledge base to implement them, nor the distributional problems that are raised, will be sufficiently small that they can be ignored.

References

Baumol, W.J. and **Oates, W.E.** (1979) 'Designing effective environmental policy', Ch. 22 in their *Economics Environmental Policy and the Quality of Life*, Prentice-Hall, Englewood Cliffs.

Turvey, R. (1963) 'On divergences between social cost and private cost', *Economica*, NS, pp. 309–13.

Chapter 9
The contribution of economic instruments to solid waste management   *J. A. Butlin*

Introduction

The uncontrolled operation of markets has both desirable and undesirable consequences. Amongst the desirable consequences are promptness of response to demand or supply changes, due to the response of producers and consumers to information on prices. The general tendency of an 'ideal-state' economy (with no economic friction resulting from imperfect knowledge, or excess power, on behalf of any of the participants in any markets) to promulgate technical and economic efficiency is a further advantage of the free-market system. A third major advantage is that the responsiveness of producers and consumers to prices (that is, the fact that they do respond, rather than the magnitude of the response) coupled with the tendency to economic efficiency, serves to allocate the limited resources (of capital, natural resources and manpower) available to satisfy to the greatest extent the largest number of wants within the economy, given the distribution of wealth.

The rationale for intervention in such a system, that is for some form of imposed control, is predicated on the concern that the benefits from the 'ideal-state' economy could not be achieved, without some form of regulation, because:
1. The ideal-state economy did not exist, some participants having imperfect knowledge about the transactions being undertaken, or some participants having more influence in the market than others;
2. For some goods and services (and 'bads' and 'disservices') generated as a result of economic activity, there were no markets through which recipients and producers could interact;
3. More recently concern over using purchasing power as a means to weight the votes of consumers of goods and services (that is, concern over the distribution of wealth and income) led to intervention in the market to try to reduce disparities in income distribution.

The purpose of the chapter is to examine the role of economic incentive systems as waste management policy tools. instruments applicable to particular products, such as deposit/refund systems, are not considered here.

Government intervention and waste management

Any intervention in a market, or the means whereby a government intervenes in a market, may be thought of as the use of an economic instrument. Economic instruments under this definition, however, would cover almost all central authority interventions.

Interventions may be thought of as either directed towards prices or quantities. In all but the most extreme market situation[1] intervention directed at one variable must influence the other (as well as prices and quantities in related markets). It is more usual to regard as economic instruments policies whose primary target is the price of a particular good or service. This can be achieved either directly, using taxes or subsidies directed at the producer or consumer, or indirectly, by subsidising or taxing products produced or purchased by producers or consumers. Examples of direct taxes or subsidies would be those of a higher rate of corporation tax for producers not using a certain proportion of recycled materials, or allowing consumers to set repair and maintenance expenditures on durable goods against their personal tax liability (to encourage greater longevity for consumer durables and automobiles). Examples of indirect instruments would be excise taxes imposed on packaging to reflect the full waste collection and disposal cost, or *ad valorem* or per unit subsidies placed on recycled materials.

Markets for certain goods do not work adequately because the goods are so-called 'public' goods (mainly because, by their nature, individuals cannot be excluded from enjoying benefits from them). Many environmental services are of this nature, and are therefore produced either directly by governments, or by governments and private enterprise together. In several countries, the provision of amenity services is achieved in this way. The provision of 'public' environmental goods and services may be thought of as the use of economic instruments.

Finally, markets may be seen not to operate due to lack of information, either on the part of producers or consumers. The success of waste exchanges in several countries exemplifies how government provision of information on the availability of waste materials can be instrumental in establishing markets that had not previously existed, by bringing together (or at least bringing to light the existence of) supplies of the demands for particular waste products.

In addition to policies whose prime target is an environmentally-based objective, there are other policies which, by their nature, have direct implications for solid waste management or resource recovery. Two examples of the many available are: the tax treatment of virgin materials producers; and the demand by central authorities for factors of production (known in some countries as governmental procurement requirements). Any comprehensive evaluation of economic instruments ought to take into account at least the most obvious of these.

In summary, to delimit the scope of the discussion, we shall adopt the following useful, if rather broad, definition of an economic instrument: any government policy which aims to influence the costs of providing or maintaining environmental goods or services; or aims to use government expenditure to supplement private production of goods and services; or aims to provide a means whereby markets can be created. A rather crude generalisation of this would include all policies, relating to solid waste management, which involve the taxes, subsidies or other forms of government expenditure, as economic instruments. Policies which are directed solely towards quantities, and are implemented mainly by specific legislation, may be thought of as regulatory (although they obviously have economic implications, as was noted above).

The role of economic instruments in solid waste management policy

The role of economic instruments in solid waste management can best be seen in the context of the policy process as a whole. Theil (1958) has provided the general outline of steps in the policy process. He envisaged the policy process as one whereby general policy goals are achieved using economic instruments to affect (or 'impact on') particular or specific targets. For example, the general goals of reduced solid waste disposal and improved resource recovery can be achieved, *inter alia*, by a specific policy directed towards particular items entering the solid waste stream, such as beverage containers and packaging. The proposed US product charge legislation was a useful example (US EPA, 1977). As particular targets are approached through the operation of the economic instrument, so are the more general goals of improved solid waste management and resource recovery approached. Thus, economic instruments should not be viewed in isolation, as ends in themselves, but rather in the context of the targets which they are intended to achieve, that is, as a means to an end.

From the foregoing it is apparent that economic instruments need to be assessed in the light of possible policy targets which they might be used to achieve. To some extent this will reflect past experience, and to some extent it will reflect experience with similar economic instruments in different policy spheres.

Categories of economic instruments

The placing of policy instruments into categories will inevitably lead to the problem of particular instruments not fitting into any particular category, or of certain instruments being eligible for more than one category. Categorisation does, however, provide an organisational framework, and it is for this reason it is resorted to here.

There seem to be at least two eligible ways in which economic instruments used in environmental management can be categorised. One way is by that part of the environmental degradation towards which they are directed, that is: air, water, noise, and solid waste. The major disadvantage here would be in the number of instruments which would overlap categories. The major alternative basis for categories is by the purpose for which economic instruments are intended: to internalise external environmental costs, to provide public environmental goods and services, to provide adjustment aid for increased resource recovery, or to assist in the establishment of secondary materials markets and trade in recovered materials.

Whilst this second basis for grouping instruments will have overlaps they will be considerably less, and, it is felt, will assist in understanding the use of particular policy instruments and the circumstances under which their use has been found to be, or is most likely to be, effective. Before a full discussion of the instruments, however, another issue must be resolved.

Economic instruments: nature, purpose and potential

The various classes of economic instruments that constitute the range of instruments under review have been outlined above. They broadly represent policies to correct externalities; policies to complement the supply of environmental goods and services with strong 'public goods' attributes; and policies to encourage the establishment of markets, the costs of establishing which would otherwise be prohibitive in relation to the individual benefits that could be realised. Most attention has been directed to the first category of instruments, but this does not detract from the importance, in practice or potentially, of the other two classes of instruments.

Charges and solid waste management

The use of economic instruments, as opposed to a regulatory approach, in solid waste management amongst developed countries varies from the indirect provision of waste collection, to the direct charge for municipal waste collection (user charges), to a charge on specific categories of waste and further, to the payment of a bounty on the return of specific categories of solid waste to specific collection centres. These policies are mostly incentive-based.

User charges

A user charge is essentially a payment for household waste disposal based upon the approximate weight or volume of solid waste. This simple definition conceals a large number of possible permutations between the charge base, the extent to which the waste must be separated by the household, and whether or not the charge could be paid directly or deducted from a property tax base. The administrative issues notwithstanding, the important question is whether the amount of household wastes generated are at all sensitive to the costs of collection and disposal.

The question of the basis for a user charge for municipal waste disposal services relates to other charges as well (the product charge particularly, to be discussed below). It is, therefore, appropriate to discuss this matter here.

Possible material characteristics to be considered as the basis for solid waste charges include weight, volume, compacted volume, weight of incinerated residue, and factors relating to the ease of recycling and disposal (US EPA, 1977). The problem with mixed municipal wastes is that, in principle, the charge should be based upon the collection and disposal costs of the components. In practice, however, such an approach would be totally impractical. As has been mentioned above, both weight and volume are convenient bases for the charge. Both, however, give an incentive to avoid the charge (even though this may only marginally detract from the waste reducing effects of the charge).

Deciding upon the appropriate rate for the charge will also provide problems. The charge should cover the full marginal costs of collection. However, recent work in the United States raises a difficulty even at this stage. Attempts to estimate the cost curves for municipal waste collection and disposal both by public and private authorities revealed falling marginal and average social cost curves. The equating of the charge with the full marginal costs of collection and disposal is predicated upon rising marginal and average costs. If costs are falling, that is if waste collection and disposal are characterised by significant economies of scale, then equating the charge with the marginal costs of collection and disposal will not cover the full costs. The charge would need to be established at the average cost of collection and disposal.

Product charges[2]

The concept of including in the final cost of a good a charge to cover the collection and disposal of the discarded good, or of the associated packaging, is obviously founded firmly on the Polluter Pays Principle. The product charge has been defined as:

> an excise tax on the material content of consumer products entering the solid waste stream.

The key points relating to the principle of the product charge are:

1. The products to be included;

2. Provisions for exempting reclaimed and recycled materials used in the products or in the packaging;

3. The extent to which such an indirect tax will be regressive, i.e. the extent to which the tax will fall more heavily on the shoulders of the poorer members of society.

In principle the product charge should cover all products entering the municipal waste stream from domestic consumption. In practice, however, it is felt, at least in the United States, that the administration of a product charge on all goods would be too complex and too costly, and that the charge should be directed at products which comprise a large proportion of the household solid waste generated. The proposal in that country was to impose the charge on paper products and non-paper packaging materials. These two categories together constitute an estimated 80 per cent of the United States solid waste stream. Of the possible bases for the charge, that based on weight seemed to be the most suitable. The charge should be set at marginal direct cost of collection and disposal and the charge should be imposed as near to the point of manufacture as possible, to reduce the number of points at which the charge should be collected, and where regular monitoring needs to be carried out.

The recycled material content of goods and packaging is another problem associated with the product charge. The problem is essentially one of identifying the amount of material reclaimed from household solid waste that is included in the manufactured product, and of assessing the value of this. ('Prompt' industrial scrap should not be included in the recycling credit.) The imposition of the tax or charge near to the point of manufacture considerably reduces the problem of assessing the reclaimed material content of goods or packaging.

The question of whether this charge, like any other indirect tax may weigh more heavily on the poorer members of society, is a matter of concern. It is popularly believed that any indirect sales tax will be regressive. However, in discussing a similar issue, a recent report came to the following conclusion:

> In analysing the effect of /a/ new tax on the distribution of income, it is important to recall that a single component of the government budget constraint cannot be changed in isolation. If tax revenues are increased by the institution of a new . . . tax, then there must be a counterpart change in government spending on goods and services or on transfer payments and subsidies, in monetary policy, in debt management policy, or in the levels of other tax revenues . . . the important point is that the counterpart change could well take the form of alterations in other taxes so that the combined effect of the measures would be distributionally neutral. The significance of this point is that it provides an immediate answer to the common, but misleading, complaint that sales taxation of any kind is regressive, and therefore to be avoided. It is far from certain that /such a/ tax . . . would be regressive, but in any case any such effect could be offset (Butlin and Sumner, 1978, p.50).

A further concern about the product charge concept relates to the elasticity of demand for waste disposal services. If the elasticity proved to be low the effect of a product charge could be expected to be relatively insignificant.

Conclusions on the use of economic instruments in the field of domestic solid waste management

The user charge is used to a lesser extent than might be expected, given its relative ease of application when based on a charge per standard container system. Under such a system, the charge should optimally be the full marginal cost of collection and disposal per container (in the absence of significant economies of scale in either collection or disposal). The charge revenue can be used to finance municipal collection, disposal and resource recovery facilities, or can be redistributed to assist private facilities. The administrative costs are likely to range from 2 per cent to 5 per cent of the total cost of the scheme. In practice, the charge does not appear to be related to the full direct and indirect costs of collection and disposal.

The major barrier to the introduction of a user charge system in more countries would appear to be the system of joint financing of municipal refuse collection by local authorities and by the national government. It may be argued that households would pay several times over for solid waste collection: first, from the taxes used to pay the national contribution to municipal collection; second, from local property taxes; and third, from the user charge. There is, however, no reason why the charge should not replace the municipal property tax contribution to municipal waste disposal, or, indeed, the whole cost of municipal waste disposal, avoiding the double transfer of funds from taxpayer to central government to local government.

The user charge approach appears to offer the basis for a simple, comprehensive municipal waste collection fee. It does not require particular regulatory or administrative instruments to facilitate its implementation, and would not conflict with other, specific instruments designed to cope with other, specific categories of waste.

The product charge is also an instrument which offers relative ease of application with a comprehensive coverage, even if only directed at specific categories of waste, as in the United States. If imposed on a weight basis for all flexible packaging, a per unit basis for rigid containers, tyres and bulky consumer durables, it would cover a large proportion of the products that ultimately enter the solid waste stream. Being imposed as near as possible to the manufacturing stage would reduce collection and administration costs, even though there may be some danger of percentage mark-ups increasing the cost to the final consumer.

The product charge, like the user charge, is a comprehensive solid waste management policy instrument. It is of little use trying to decide which is the 'better' policy. Each has its relative merits, and either may be the more suitable in a particular situation. The advantages of the user charge have been outlined above. The product charge has the following advantages:
1. Being imposed at the manufacturing stage, there is a strong incentive for products to be redesigned to be less solid-waste-intensive.
2. The system of recycling or reclaimed materials credits acts as an incentive for resource recovery and the use of reclaimed materials in manufacture;
3. Like the user charge, the redistribution of revenue from the charge will aid both solid waste disposal and resource recovery at the local authority level;
4. Like the user charge, providing statistics on waste collection and disposal are available, the establishment and adjustment of the charge according to satisfactory bases is not difficult;
5. Like the user charge, the product charge does not require specific administrative and regulatory support (other than that required to establish the charge system), nor does it conflict with other economic, regulatory or administrative instruments that may be implemented to improve particular aspects of solid waste management.

In other words, the user charge and the product charge can be seen as useful, operational tools in the solid waste manager's workbox. They are not the answer to all solid waste management problems, but if he ignores them he significantly reduces the range of policy options at his disposal.

Postscript

A US report (Resource Conservation Committee, 1979) has recently quashed proposals to introduce a product charge to aid waste management in that country, but has supported the introduction of other economic instruments. The Resource Conservation Committee advocated the introduction of, local user charges (defined as variable fees that change according to the quantity of waste to be collected). In contrast they found against, *inter alia*:

1. The immediate removal of federal tax subsidies on virgin materials;
2. The implementation of extraction taxes (sometimes called 'severance' taxes) on virgin materials;
3. The introduction of subsidies to increase the rate of resource recovery;
4. The introduction of a national litter tax;
5. The introduction of a national solid waste disposal charge, or product charge.

Amongst the reasons given in the report for rejecting the product charge are the following:

1. Data, particularly the marginal costs of solid waste collection, transport and disposal, and the elasticities of demand for the products which would be affected, are uncertain;
2. Differences in solid waste management costs between communities lowers the benefits from implementing a single, uniform national charge rate;
3. The design of a disposal charge to achieve the theoretical benefits of such an instrument would be difficult in practice.

The use of economic instruments as policy tools to improve solid waste management is also increasing in Europe, particularly in Germany, France, Norway and Sweden. In almost all cases, the instruments are directed towards particular products, and in many cases take the form of deposit/refund systems.

The evolution of environmental policies is showing that economic instruments have pragmatic as well as theoretical contributions to make to waste management. Consistent among the benefits attributable to the use of economic instruments are the achievement of particular policy targets at the lowest social cost (compared to either regulatory measures or the use of suasion). Other advantages attributed to economic instruments in this context include: the comprehensiveness of coverage; and the difficulty of avoidance by extensive court proceedings. With increasing concern about the economic inefficiency caused by bureaucratic intervention in the operation of the economy, the use of economic instruments in solid waste management can be expected to increase in the future.

Notes

1   That is, where either demand or supply are totally unresponsive to price changes.
2   Much of the information contained in the following section comes from US EPA (1977) or studies referenced in this.

References

Butlin, J.A. and **Sumner, M.T.** (1978) *The introduction of a Tax with a View to Reducing the Consumption of Non-renewable Resources*, a report prepared for the Environmental and Consumer Protection Service, Commission of the European Community.

OECD (1980) *Pollution Charges: An Assessment* (2nd Edition) Environment Directorate, OECD, Paris.

Theil, H. (1958) *Economic Analysis and Policy*, North Holland, Amsterdam. Resource Conservation Committee (1979) *Choices for Conservation*, US EPA SW–779, Cincinnati.

US EPA (1977) *Resource Recovery and Conservation*, Fourth Report to Congress, US Environmental Protection Agency, Washington DC, Appendix B, 'The status of product charge studies', pp. 88–99.

Chapter 10

An economist's view of pollution charges as regulatory instruments   *J. M. Marquand*

The purpose of this paper is to show why indeed it is foolish of any economist to claim that there is a simple answer of uniform application to the question as to whether charges for discharges into water courses are to be recommended as an efficient and equitable means of attaining desired water quality objectives.

Let us start by considering a crude version of the economists' case for arguing that charges for discharges are desirable. Pollution causes damage, and each additional unit of pollution (however measured) causes additional damage (again, assuming that some appropriate way of expressing this damage, perhaps in monetary terms, can be found): Assume then that we can draw a curve showing the additional damage caused by additional units of pollution from a given discharge. We will call this curve the marginal damage cost curve. Similarly, we can in principle draw a curve relating the extra costs to the firm of abating an additional unit of pollution to the total amount of pollution (or volume of discharge given concentration) which it emits. We can call this curve the marginal control cost curve. Where the two curves intersect, the costs of abating an additional unit of pollution equal the damage caused by that unit of pollution. In Fig.10.1, to the left of the point of intersection, the costs of abatement exceed the damage caused by each unit of pollution; to the right of the point of intersection the damage is greater than the marginal cost of abatement. Accordingly, in some sense the costs to society will be minimised if the firm abates its pollution to the point where the two curves intersect. To the left of the point, the costs to the firm outweigh the benefit to society of the abatement; to the right the costs to society of the pollution outweigh the costs to the firm of not polluting. There are in principle two ways of ensuring that the firm discharges pollutants only to the extent indicated by the intersection of the two curves. Either it can be told simply that this is the maximum which it may discharge, or it can be made to pay a charge per unit of pollution emitted of an amount OC (shown in Fig.10.1), such that it prefers to abate its pollution as far as the point of intersection rather than pay a charge, because its abatement costs are lower than the charge per unit.

Why then do economists often argue that a system of charges is to be preferred to a system of standards? The argument is usually couched in terms of the lack of information available to the standard-setting body. Competent authorities, such as Water Authorities, usually have knowledge of the condition of water courses and of the discharges to water courses, but they do

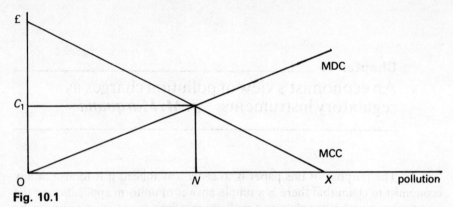

Fig. 10.1

MCC = marginal pollution control cost curve
MDC = marginal damage cost curve
When a charge OC per unit of pollution is imposed, the firm abates the
pollution it emits from OX (in the absence of any charge) to ON

not have knowledge of the costs to the individual dischargers of varying their
discharges. Figure 10.1 shows the decision problem facing only one firm.
Where there are several dischargers, the costs to the different firms will
probably vary. The least-cost way of achieving a reduction in the pollution
emitted, to a desired total, is to allow each firm to choose how much it will
pollute, faced with its own pollution abatement cost curve and the common
damage cost function. If it is faced with a charge per unit of pollution intended
to simulate the marginal damage costs to society if pollution were to be abated
to the target level, then each firm can make its own decision about the way in
which it will abate its own pollution. The Water Authority need only fix a single
per unit charge for each particular pollutant which it is considering in this way;
it does not have to set individual standards for each separate discharger. It is
argued that in setting standards, if the costs to a firm are not fully known, the
Water Authority may set standards which are unduly harsh for some firms and
unduly lenient for others. If it is determined to ensure that no more than a
certain total of pollution is emitted, it may set standards which are in general
too tough and hence cause a misallocation of resources to pollution abatement,
which would be better spent on other things.

It is sometimes objected that as much knowledge is needed to set
charges sensibly as to set standards. The proponents of charging would argue,
quite correctly, that you do not have to get your charges quite right in the first
instance. At first you set what you think will be the appropriate charge. If it is
too high, more abatement will take place than intended and in a subsequent
period the charge can be reduced. If it is too low, less abatement than desired
will take place. In subsequent periods the charge will have to be raised until the
desired water quality is achieved. Provided sufficient precautions are taken to
restrict discharges of toxic substances or of substances which inhibit treatment
of water to a desired standard further downstream, there is no reason to
criticise proponents of charging for pointing out that iterative procedures may
sometimes be needed before the appropriate level of charges is reached. The

arguments which restrict the feasibility of charging schemes are more fundamental than this.

There is also nothing wrong with the crude theory set out above as a piece of abstract theory relating only to efficient allocation of resources. If we wish to question it, or to examine the limitations surrounding its application, we must examine carefully the assumptions which are required in order for it to hold.

Note first that there is no need to use a system of charge if there is only one polluter, since it is easier to set a standard in such a case in order to achieve the desired quality of the water course. In the case where all polluters are similar, in the sense of emitting the same pollutants in similar quantities and faced with similar control costs, it is likewise unnecessary to consider a charging system, since the permitted amount of discharge compatible with maintaining the desired quality of the water course can simply be allocated equally between all the polluters. The nearer the observed situation is to either of these cases, the weaker the case for a system of charges. It is where there are many polluters using different processes and hence faced with different control cost curves that the c for charges as conventionally presented is at its strongest.

Note also that it is not necessary for the marginal damage cost curve to be known for a charging system to be used, or indeed, for standards to be set. Figure 10.2 shows that for a series of firms with different marginal cost control curves, we can either set standards (ON_1, ON_2, ON_3 respectively) or set a uniform charge per unit of pollution discharged (OC_1) in order to attain abatement of pollution to a given specific total ($ON_1 + ON_2 + ON_3$). In the case shown, with a uniform charge, the polluters abate their pollution to differing extents from the amounts (OX_1, OX_2, OX_3) which would have been discharged if there had been no restrictions. In this case the additional cost of abating one more unit of pollution would be the same for each firm. If standards were set allowing discharges other than the quantities shown in Fig. 10.2, some polluters would be abating units of pollution at higher cost than would be necessary if the quantity they were allowed to discharge slightly and

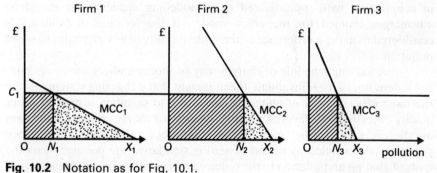

Fig. 10.2   Notation as for Fig. 10.1.

░░░ Total pollution control costs incurred by each firm i in abating pollution from OXi to ONi

▨▨ Total charge each firm pays on its residual pollution ONi when a charge OC_1 is levied on each unit of pollution emitted.

the quantity of other polluters' discharges were reduced. A uniform charge in such circumstances minimises the abatement cost to polluters as a whole if pollution is to be reduced to the desired extent, although in addition each firm makes a payment to the charging authority on those units of pollution which it chooses not to abate.

Note also that there is no problem in using a system of charges in conjunction with a system of maximum permitted discharges for particularly noxious pollutants. There is no reason to fear that a charging system used in conjunction with such regulations will lead to dangerous or highly undesirable levels of pollution any more than a system based on regulation alone. There is also no reason to think that such a system will be less effective *qua* charging system than one which tries to operate without supporting regulations.

However, it has been assumed for the purpose of the analysis so far that units of pollution, however measured, are equally undesirable wherever they are emitted. Such an assumption ignores variations in the assimilative capacity of the receiving waters. To ignore such a valuable environmental resource is strange when we are concerned with the efficient allocation of resources to achieve a desired environmental quality.

Once it is recognised that assimilative capacity varies from one place to another, the undesirability of a given concentration and quantity of a pollutant varies accordingly. The abatement we require from particular polluters in order to achieve a given environmental quality most efficiently will thus vary as well. If we apply sophisticated river modelling to handle such cases, we will find that a variation in charges from one polluter to another, according to the assimilative capacity of the relevant receiving waters, emerges as the most efficient solution. Only when all the polluters concerned can be regarded as though they were in the same location, can a uniform set of changes be used to achieve efficient allocation of discharges between polluters. Thus the crude model may be applicable where polluters discharge into an estuary, or where a river can be divided into zones within which the polluters can be considered to occupy the same location. Only a much more complicated system of charging, in conjunction with sophisticated river modelling, achieves the economic advantages claimed for the crude model, if the location of polluters is considered to make a difference to the undesirability of any given discharge of pollution.

A varying schedule of charges may be more cumbersome to operate, but it does not of itself invalidate the argument that a charging system may be the most efficient way of allocating resources to maintain a desired water quality objective. However, a varying schedule of charges raises a significant question of equity rather more acutely than does a uniform charge. When pollution is controlled by means of setting standards, the polluter bears the costs of abating his pollution to the required extent but incurs no other costs. If pollution is controlled by levying a charge per unit of pollution discharged on all units of pollution, then the firm bears not only the costs of abating pollution to whatever extent it decides to do so, but also pays a charge on the residual units of pollution. Such a charge is in many ways akin to a tax; it is not used to purchase services of any sort. Now there is no reason in principle why Government should not levy taxes on the use of the environment in this way. No Government in fact does so. In cases where systems of pollution charging have been considered, it has often been suggested that the charges be paid to the regulatory authorities rather than to central or local Government. But to the extent that such a system of charges is akin to a system of taxes, it would seem inappropriate for the revenue to accrue to anything other than an authority such as central Government which has the explicit power to levy taxes. Moreover, in the case which we discussed, where the location of a polluter, even within one river basin, influenced the charge per unit of pollution which he paid, a situation would arise which is contrary to normal fiscal practice, of discrimination between taxpayers for reasons which might prove difficult to set out in a general enough form to prove acceptable.

Although there is no reason in principle why taxes should not be levied on the use of the environment, the fact that no Government has such a system of taxation in practice makes it contrary to the interest of any one country to impose such a system of taxation unilaterally unless it sees some environmental benefit accruing sufficient to offset the adverse effects upon trade. Yet in principle it is possible to achieve the same environmental effects without any net revenue accruing to the charging authority and hence without imposing undue costs on industry. Indeed, it may be possible to arrange matters so that polluters collectively bear lower abatement costs than those shown in Fig. 10.2. Whilst the possibility of efficient schemes without net revenue exists, the deep founded resistance from industry towards charging schemes which produce net revenue is not only understandable but entirely rational.

We shall call schemes where there is no net revenue 'redistributive charging schemes'. Under such schemes, polluters make payments to an agency, which either undertakes pollution abatement itself or pays other firms to abate pollution more than they would otherwise have done. The first of these possibilities is of course a familiar one where aspects of water services other than direct discharges to water courses are concerned. There is no reason in principle why water authorities should not construct sewage treatment plants of their own for treating discharges which would otherwise have gone direct to water courses, and no reason why, in such circumstances, the polluter should not pay the costs involved to the water authority. But some countries have experimented with redistributive schemes which are rather different from this. The French Agences de Bassin are the nearest examples. The Agences collect payments from some firms and transfer them to others to pay for pollution abatement equipment. Thus those firms where there is the technical possibility of cleaning up relatively cheaply undertake the cleaning up for everybody (always assuming that there are no locational problems). The French and the Dutch too appear well satisfied with their redistributive schemes. There is thus no doubt that such schemes can operate effectively in practice. Such schemes are also in accordance with the 'polluter pays principle' as set out both by OECD and by the EEC, so long as the redistributed funds are indeed used entirely for pollution abatement purposes. What requires to be considered then is under what conditions such a scheme is preferable to a system of pollution control based entirely upon standard-setting.

There are some practical problems in applying a redistributive scheme in an efficient manner. Consider the firms shown in Fig.10.2. The costs which each bears under a charging scheme with no redistribution are shown by the dotted triangle, representing the pollution control costs incurred by each firm, plus the shaded rectangle showing the total charge each firm pays for its residual pollution which it decides not to abate. Our problem is to find the charging system which will produce the same amount of abatement on balance, without making firms pay also the charge on the residual damage. The means of doing this is shown in Fig.10.3.[1] A charge per unit of pollution is imposed which

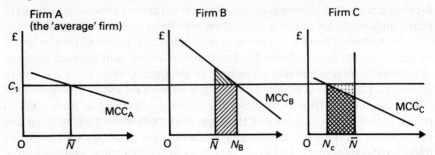

Fig. 10.3 Notation as for Fig. 1.

Firm A is the 'average' firm and neither pays a charge nor receives a payment to abate pollution further.

Firm B pollutes more than the 'average' firm and pays the agency ▨ for $\bar{N}N_B$ units of pollution at charge OC_1 per unit. It would be prepared to pay ▨ in addition rather than abate its pollution to $\bar{N}$.

Firm C abates further than the 'average' and receives a payment from the agency to compensate it for so doing. The minimum payment required to induce firm C to abate from ON to ON_C is ▨. However, the agency may be prepared to pay ▨ in addition, since it would then be paying OC_1 per unit of additional abatement, which is the payment required if the agency is to receive no net revenue.

is no different from the charge per unit imposed on the firms in Fig.10.2. However, the charge is levied only on units of pollution beyond a certain number. The number of units of pollution beyond which a charge is levied is determined as in some sense an 'average' amount of pollution emitted, in circumstances where the total amount of pollution is at an acceptable level. If a charge (at the same level per unit as in Fig.10.2) is levied on units of pollution emitted beyond the 'average' number and if this revenue is redistributed to firms who emit less than the 'average' amount of pollution and used by them towards their purchase of pollution abatement equipment, then all the requisite conditions for the efficient operation of the scheme will be satisfied.

A point to note is that the firms which make payments rather than abate their pollution are, by definition, prepared to pay more per unit of pollution than the cost per unit to the firms who undertake the extra abatement. The area under MCC_B between $\bar{N}$ and N_B in Fig. 10.3B is greater than the shaded area representing a charge OC_1 per unit of pollution, and the rectangle with side $N\,\bar{N}$ in Fig. 10.3C, representing a payment to firm C of OC per unit of pollution abated from $\bar{N}$ to N, is in turn greater than the hatched area under MCC. At a charge and a payment of OC, per unit of pollution, the abaters make a small profit (marked with crosses in Fig.10.3C) on their extra abatement activities and polluters retain a small benefit (the triangle shaded from left to right in Fig.10.3B) from polluting and paying rather than abating. Alternatively, the agency could charge polluters slightly more and/or pay abaters slightly less, thus receiving a certain revenue. It may be in practice that this is used to cover the agency administrative costs, for example, but none the less it is worth noting that there is no unique answer to how much a given set of polluters should pay and how much a given set of abaters should receive in order that the desired abatement be achieved at least cost to society. Someone stands to make a net gain from the redistributive scheme under almost any feasible set of arrangements.

Additionally a host of practical problems arises. On what basis is the 'average' amount of pollution estimated? Is it units of pollution of whatever sort concerns us, or units of pollution per unit of output, or units of pollution on some other basis? Even if this problem is overcome, other questions remain.

Many of the other questions are associated with the set of assumptions which we have not yet questioned. We have assumed so far that it is straightforward to draw a marginal pollution control cost curve for a firm, that such a curve is known at any one time and that such a curve is continuous. In practice this if often not the case. All that we can be certain about is the expenditure which a firm is actually incurring which is directly attributable to pollution control. Thus we can be sure of observing one small stretch of the curve in the immediate vicinity of the present situation. But other points on the curve are far less certain. It must be noted that there are two main ways of altering the amount of pollution which a given firm emits. One method is to add on specific pollution abatement processes to whatever processes the firm is already undertaking. Such additional processes are usually relatively easy to cost, but they are likely to come in discrete lumps so that the curve is not necessarily continuous. The other way in which a firm can change the amount of pollution it emits is by a fundamental change in its production process, so that a different collection and a different quantity of pollutants are produced. In such a case it may not be at all clear what costs should properly be attributed to any additional pollution abatement. Accordingly the nature of an appropriate payment to compensate a firm for undertaking additional pollution abatement in this manner is extremely difficult to define. And certainly where changes in production processes are in prospect, there is no guarantee at all that there will be a continuous marginal control cost curve; major discontinuities are probable. Thus we do not in general have a smooth curve but only a series of identifiable points which show what the effect of the different pollution abatement options for the particular firm would be. A firm may incur very substantial expenditure to move from one point to another. It will be impossible for it to deliver minor adjustments in the amount of pollution it emits in exchange for relatively small payments.

We have seen that, where it is impossible to determine what costs are properly attributable to additional control of pollution, it is impossible to determine what payment from the redistributive agency should properly be made in order to compensate the firm for the additional pollution abatement which it has undertaken. Thus we have reached a situation where it is very hard to distinguish between payments used to make sure that pollution abatement expenditure is undertaken which would not have been undertaken otherwise, and payments which are just straight subsidies.

In addition it should be remembered that the operation of the whole system presupposes that differences of location are such that abatement by one polluter can be replaced by abatement by another.

It should thus be apparent that it is almost impossible to appraise whether a particular redistributive scheme operates more efficiently than a scheme of standard-setting to achieve corresponding results. Against the system of standard-setting it should be said that sometimes there may be a choice between imposing extremely stringent standards on one particular firm with low abatement costs and less stringent standards on firms with higher abatement costs, but that this option is not available in practice because there is no way of spreading the cost burden between the firms involved.

So the proof of the pudding must lie in the eating. Either system can operate well as a matter of administrative practice. Both systems are prone to inefficiency or inequity under particular circumstances. Whichever system is used, redistributive charging to supplement standard-setting, or standard-setting on its own, the agency operating the scheme needs to be aware of the circumstances where the scheme may lead to undesirable results, and needs to conduct particular studies from time to time to make sure that such results are not occurring on any significant scale.

And when all is said and done the presumption must be that charging tends to become more appropriate than standard-setting alone the more that there are many disparate dischargers rather than when there are few; that charging schemes tend to become either complicated or inefficient where polluters cannot be regarded as discharging into the same body of water, and that charging schemes for direct discharges into water can and should only be operated in conjunction with standard-setting where the more unpleasant discharges are concerned. Charging schemes for direct discharge may be helpful where there are many dischargers into an estuary; they are unlikely to be helpful when the direct dischargers are spread out along a fast-flowing river. Each case should be judged on its merits and the judging will not always be easy.

Note

1   See 'Note on some aspects of the Polluter Pays Principle and its implementation' by Judith Marquand and David R. Allen in *The Polluter Pays Principle: Definition, Analysis and Implementation*, OECD, 1975, especially pp. 85–7 and diagrams on p.92. The system described here is derived from, but not identical with, the scheme set out there.

Part Five

Look round the habitable world! how few
Know their own good; or knowing it, pursue.
Dryden (translating Juvenal)
International aspects of environmental management

Introduction _____

The international dimension of the management of the environment is one to which the student of environmental economics is rarely exposed, and yet it is one in which there is a great deal of activity. There is great concern about the transport of SO_x from north-west Europe to Scandinavia; about pollution from oil tankers; and about the global degradation of the ozone layer in the stratosphere. At the international level the political dimension of environmental management is even more clear than in national environmental management problems, making the achievement of an environmentally satisfactory solution (that is, a solution that will not lead to irreversible environmental degradation) even more difficult.

The two authors whose contributions complete this volume are eminent in the field of international environmental management. Professor Walter's concern is with, *inter alia*, the effect of national environmental policies on the qualitative impact of environmental policy on international trade. His paper is a thoroughgoing summary of work in this area.[1]

Michel Potier takes a more pragmatic view of the international problems of environmental management. His experience with OECD's Environment Directorate qualifies him well to examine the role of inter-governmental international organisations in the field of environmental management. In the course of his paper he reveals a level of international activity that is known to relatively few people but through which many of the attempts to agree on international management issues are pursued. It is appropriate that this text, designed to give the reader a feeling for the link between environmental natural resource economics and policy issues, should end with such an explicit orientation towards environmental policy.

Note

1 For a more detailed discussion see his *International Economics of Pollution*, Macmillan, London, 1975.

Chapter 11

A survey of international economic repercussions of environmental policy
Ingo Walter

Environment is one of a number of areas of public-policy concern where recent developments have been rather rapid. It is also an area – like consumer protection and occupational safety and health – where compliance with new governmental regulations has fundamentally affected the economics of production and consumption. As in these other fields, environmental policy is essentially a national issue, because environmental problems that are directly international in character – e.g. atmospheric and ocean pollution – comprise a relatively small part of the whole, and because institutions that might address themselves effectively to transfrontier questions are only now beginning to emerge. The fact of national sovereignty in environmental policy, when coupled with its economic consequences, leads directly to repercussions on international economic relations.

This paper reviews the sources of environmental diversity among countries, the importance of environmental cost-allocation principles, the implications for trade flows of both production- and product-pollution control measures, and the significance of environment-induced relocations of productive activities.[1]

International variations in the framework of environmental policy

Measures designed to maintain and restore the quality of the natural environment appear to be subject to substantial variation among countries. In some, strict pollution control legislation has been enacted, coupled with the administrative machinery designed to enforce compliance. In others, pollution control has been outweighed by other economic and social priorities, particularly in the developing countries.[2] Moreover, among countries that have implemented environmental controls the approaches used have tended to differ, and this has added another dimension of inter-country diversity in environmental policy. The fact that different countries take up the task of environmental management according to different schedules, with different degrees of rigour, and using different policy approaches, raises the likelihood that international trade, commercial policies, and the location of industries will be affected.

Most environmental problems can be traced to economic activities involving extraction, production, distribution, consumption and disposal of goods and services – activities that either directly, or indirectly through the generation of residuals, change natural conditions.[3] These typically consist of what economists call 'common property resources'. They cannot, or can only imperfectly, be reduced to private ownership. They can however be viewed as asset-like stocks which produce streams of varied and useful services: intangible (scenic views and recreational amenities), functional (waste assimilation, nutrient recycling, pest control, and climate regulation), and tangible (flows of life-sustaining material resources, such as water, air and minerals).

All natural systems possess a capacity to assimilate, to some degree, a broad range of residuals through natural mechanisms of transport, transformation and storage. Economic activities may diminish the capacity of natural systems to perform their waste disposal and other vital roles, or the volume of residuals generated may exceed that capacity. The significance of such modifications in human terms, however, will depend heavily on interactions of these changes with human conditions. For example, deterioration of air quality may be more serious if it occurs in densely populated rather than sparsely populated regions. Likewise, soil erosion will have a greater impact if it occurs in a region where arable land is in short rather than abundant supply (c.f. Gladwin and Walter, 1977).

The character of environmental problems within any given geographic area thus depends on (a) the changes in natural conditions and their interactions with (b) human conditions induced by (c) economic activity. Given existing inter-country variation in each of these dimensions, the resulting characteristics of environmental problems found within different nations exhibit correspondingly wide variation. These make themselves felt in a number of ways. Environmental problems may be reversible or irreversible, more or less certain, temporary or cumulative, independent or synergistic, and short or long in the time scale between cause and effect. The order of gravity may extend from losses of environmental amenities to problems that seriously threaten human health, genetic stock and the sustaining capacity of entire ecological systems. The areas affected by the environmental damage may be local, regional, national, multinational or global.

The translation of existing or expected environmental problems into corrective or preventive environmental policy depends heavily on social and political factors. Environmental problems within any nation are perceived, interpreted and given priorities in accordance with existing social preferences. Environmental quality is a matter of social choice – and societies may differ, quite legitimately, in their views as to what constitutes an 'acceptable' level of environmental quality. Societies afflicted with widespread malnutrition and disease, high infant mortality, low life expectancy, high illiteracy levels and endemic unemployment are not likely to place the same value on degradation of the natural environment as societies in which these kinds of problems have been overcome.

This means that identical objectively-perceived environmental damage may be accorded quite different social weightings in different countries. These relative social weightings then enter into the political process, where their transformation into policy action depends on the specific features of the national political system. Differences in interest-group pressures and in the existing political machinery will often lead to differences in the assessment and selection of feasible policy alternatives. Since policy structures among nations are both heterogeneous and imperfect, this means that identical social weightings applied to the same environmental issue in different countries may not in fact result in the same selection of policy alternatives.

What generally emerges from the political process is a framework of environmental policy – one that sets forth, explicitly or implicitly, the objectives to be sought, the approach and timetable to be followed, and organisational arrangements to be employed. The framework will reflect differences in the balance between centralised and decentralised authority, comprehensiveness or fragmentation in administrative structures, balance between executive and legislative control, relative strengths in staffing of environmental ministries, preferences for 'case-by-case' or 'blanket' policy approaches, and allocation of environmental costs to taxpayers or consumers.

Implementation of the policy framework will involve the selection, development and use of different kinds of environmental policy instruments. Nations typically will utilise some combination of prohibitions, standards, charges, tax schemes, land-use controls, clearance requirements, subsidies, hearings procedures, and the like in their attempts to achieve environmental targets. Such controls provide incentives and disincentives for decision makers at the operating level of the economy – 'carrots and sticks' that serve to guide the choice of inputs, processes, and outputs of environment-affecting activities (Barde, 1976).

What emerges in a picture of diversity. Generally, the developing countries appear to have less rigorous environmental policies than the developed countries. This is to be expected if environmental quality is in part viewed as a consumption good that is sensitive to income, so that poorer countries are able to afford lower environmental quality levels whenever this involves sacrificing other social or economic objectives. However, a number of forces are at work that may significantly reduce the extent of environmental-policy diversity over time. There is some evidence, for example, that the demand for improved environmental quality – which normally involves costs in terms of reduced availabilities of other goods and services – is a positive and elastic function of income. If this is true, then one can expect some convergence in environmental standards as levels of real *per capita* income rise.

In certain areas, such as the European Communities, attempts at regional harmonisation of member-nation environmental policies currently under way are indeed moving ahead, albeit at a rather slow pace. Other international organisations, such as the United Nations Environment Program and the Organization for Economic Co-operation and Development, are beginning to exert an influence on the setting of national environmental policy agendas. And environmental policy innovations – e.g. environmental impact statements, toxic substances control legislation, and coastal zoning acts – which emerge in one country tend to be diffused and adopted by other nations as well.

Hence, mechanisms are beginning to work in the direction of greater homogeneity in national environmental programs, at least among the advanced countries. They have not yet exerted a great deal of influence, however, and the environmental setting of the international economy remains diverse.

International differences in environmental cost allocation systems

One of the most important dimensions of the kind of international diversity in environmental policy outlined above is the allocation of pollution control costs. Several options are available here.

First, environmental control may be pursued by means of *charge strategies*. For example, a tax may be placed on producers according to the amount of gaseous pollutants emitted into the atmosphere or the amount of liquid effluents discharged into bodies of water. Economists would set such effluent charges at levels which, at the margin, would equate the monetary value of the incremental environmental damage caused with the incremental cost of reducing the form of pollution in question. Economists can show that this is the most efficient way of controlling pollution, consistent with the ambient level of environmental quality desired, because it redirects resource-use in a least-cost manner. Under this arrangement, pollution control charges would represent variable costs in the operation of the firms in question, and would influence both the prices and the outputs supplied in the market (Ayres and Kneese, 1969).

Second, a nation may decide to pursue its environmental policy by means of a *quantitative restrictions* on the permissible amount of liquid and gaseous effluents in order to maintain the desired level of ambient environmental quality. It is up to the firm to decide how to meet the standards in question, with the result that investments in pollution-control capital equipment, technology, as well as operating and monitoring costs drive up its fixed and variable cost structures. Again, this may have significant implications for the firm's prices and output, particularly in the long term.

Both of the aforementioned techniques allocate the costs of pollution control directly to the polluter. This falls under the heading of the so-called polluter pays principle (PPP), accepted by the OECD countries as an appropriate cost-allocation standard. Such costs are in the first instance met by producers who must reduce air and water pollution levels and adapt the quality of their products to environmental requirements. The costs may then be passed forward onto product prices, thus discouraging the purchase of more pollution-intensive products and encouraging the purchase of less pollution-intensive ones. Or the costs may be passed backward to factor returns, thus making the production of pollution-intensive goods relatively less attractive. The internalisation of environmental 'externalities' therefore tends to bring about the realignment of production and consumption decisions in less pollutive directions. This is viewed as a major positive aspect of the polluter pays principle.

A third alternative is for the government to defray part or all of the costs of pollution-control activities. For example, pollution-control hardware may be subsidised by cash grants from the government, by the availability of credit on concessionary terms, by tax-abatement schemes, or by allocation of pollution-control functions to regional environmental authorities at less than full cost. In such cases the polluter obviously does not pay the full cost of abatement, and therefore less of a cost burden needs to be passed onto product prices or factor returns. So there is less of an impact on costs and prices, and on the mix of production in the national economy (OECD, 1972).

Although the OECD countries have accepted the polluter pays principle, it appears that departures from that principle are very widespread indeed. In some cases these have been justified by the level of economic disruption that would be associated with full implementation of the PPP, involving regional unemployment, plant closings, and the like. Indeed, departures from the PPP are sufficiently widespread that the principle appears often to be honoured more in the breach than in the practice. As a result, it is not accurate to view the degree of rigour embodied in a nation's environmental policy as indicative of the possible international and economic repercussions. Rather such an observation would have to be modified by specific knowledge of the cost-allocation principle being followed by the nation in question and by other countries.

Also related to the question of cost allocation is the degree to which pollution control costs end up influencing product prices. It seems clear that pollution-control costs need not be passed forward onto product prices on a dollar-for-dollar basis. The extent to which such a pass-through occurs depends in large measure on the nature of competitive conditions in the industry. The less competitive the industry, the greater the pass-through is likely to be. From an international point of view, therefore, it is important to know specifically the nature of these competitive conditions on an industry-by-industry level, in order to make accurate assessments of the degree of trade-response that is likely to occur. Appendix 1 indicates the differences in the output effects under various alternative assumptions, as revealed in a recent study by Mutti and Richardson (forthcoming).

Clearly, international competitive dislocations will – at least in the short run – be worse if some countries pursue environmental standards by means of strict enforcement practices or effluent charges, while others pursue the same objectives using direct or indirect public subsidies financed out of general fiscal revenues.

As in the case of national environmental preferences and assimilative capacities, there is as yet relatively little concrete evidence regarding existing or emerging differences in national implementation patterns. Strict adherence to the polluter pays principle thus remains somewhat of an open question, and indeed the OECD programme itself provides exceptions for 'hardship cases'. Pollution taxes and related charges seem to be relatively rare at both the national and sub-national levels, in part because of political resistance to 'licences to pollute'. Little information is yet available on direct subsidisation, both regionally and nationally, such as favourable tax and depreciation treatment of environmental-control hardware, government sponsored R&D, and access by industry to public sewage-treatment facilities. The gap between the *establishment* of environmental regulations and their subsequent *enforcement* likewise appears to be relatively wide, once again subject to the absence, so far, of reliable information.

One recent study of national implementation of environmental policy by UNCTAD, for example, shows that almost all countries that have rigorous environmental policies have found it necessary to make exceptions to the polluter pays principle (UNCTAD, 1976). Most indicate adherence to the PPP as a sensible and internationally-agreed way of allocating pollution-control costs, but most have also felt the need to implement subsidies of various kinds.

A number of countries provide concessionary loans for environmental capital projects (Denmark, Finland, Japan, Luxembourg, Norway, and Spain, among others). Some countries also provide outright grants for specific pollution control projects (e.g., Denmark and Israel) or other forms of direct subsidies (e.g., Iraq, Luxembourg, the Netherlands, Norway, Spain, Sweden and the United Kingdom), while emergency aid is available to members through the European Communities. Indirect subsidisation is also available by way of government guarantees (Finland and Norway), accelerated depreciation (Finland, F.R.Germany, Norway, Singapore, Spain and the United States), tax relief (Finland, Japan, New Zealand, Norway, Spain and the United States), subsidised research and development (Finland and Iraq), as well as exemption from import duties on pollution control equipment (Israel and Japan).

As we have noted, departures from the PPP tend to alter the trade effects of pollution control. Costs need not be reflected in prices and subsidised producers do not lose in international competitiveness. As we shall indicate below, the widespread existence of such departures means that the danger of restrictive trade policies is less serious.

Quite apart from emerging international differences in environmental standards, and in their achievement, there may also be significant variations among countries in the capacity of the environment to assimilate pollutants – in climate, geography, population density, and the like. There is of course no systematic relationship or coincidence between national political frontiers and the highly specific determinants of environmental assimilative capacity. But it is hardly unreasonable to expect that such variations do exist and will continue to exist, thereby influencing the cost of meeting whatever environmental standards individual countries decide to establish. Viewing this issue from the standpoint of international trade, environmental assimilative capacity can be considered a supply factor – just as the cost of labour or capital – while environmental standards can be considered a demand factor derived from revealed social preferences. Both will influence the allocation of relative costs and prices of internationally traded products, and hence the degree to which existing competitive patterns will be affected.

Trade and competitive effects: process pollution

Despite the absense of reliable information on environmental assi-milative capacities, national environmental preferences, and their actual im-plementation and cost-allocation patterns, it is not difficult to determine at least the *direction* of their impact on international competitiveness and trade. Countries in the forefront of the movement to safeguard the environment will tend to suffer adverse competitive shifts – exports will tend to become less competitive in world markets, while imports will tend to become more com-petitive in the domestic market. Producers located in countries selecting enforcement techniques which ensure that environmental costs are indeed reflected in product prices will suffer relative to those located in countries opting for subsidisation to achieve similar goals. Pollution-intensive industries producing tradeable products will be affected in the first instance. But so will firms using the output to these industries as important raw material or intermediate inputs (Walter, 1973). For example, induced increases in (gener-ally non-traded) electric power costs or environmental control measures undertaken in the ferrous and non-ferrous metals sector will escalate costs in various manufacturing industries producing internationally-traded products.

The overall competitive impact, of course, is likely to vary dramatically by industry and by product. But even in the aggregate, international differences in environmental management-related process costs may lead to shifts in the direction of trade, the terms of trade, the product-composition of trade, as well as the balance of trade and payments. The fundamental question is how important production costs associated with environmental management really are as determinants of price-competitiveness in the international marketplace. If there are no other barriers to effective competition – as within a unified national market or a regional market such as the EEC – the impact in certain sectors may be rather pronounced, and may even be sufficient to induce a 'go-slow' approach in environmental policy.

The international economy, of course, is characterised by a variety of other competitive distortions – such as tariffs and non-tariff trade barriers – and hence the relative importance of costs induced by environmental control may be diminished accordingly (Walter, 1974). On the other hand, *within a nation* interregional variations in environmental norms and their competitive effects on trade are moderated by the existence of a strong national government charged, in principle, with minimising any disruptive economic and competi-tive impacts. Internationally, with political sovereignty much more important and in the absence of supranational authority, the competitive implications for trade flows may be both stronger and more durable.

Recent studies have attempted to establish the extent to which existing or proposed environmental standards affecting production processes might influence product costs and hence international competitiveness and trade. Using an econometric model of the US economy and projected future American environmental measures (assuming other nations do *not* follow suit), one concludes that during 1972–80 pollution control and compensatory macroeconomic policy measures are expected to exert a negative impact on the US balance of trade in the amount of $1.9 billion. If, on the other hand, pollution control costs are 50 per cent *higher* than current estimates, or standards are raised correspondingly, the average annual 1980 negative trade balance effect on the US is estimated to be in the neighbourhood of $3.2 billion or 6–7 per cent of current levels of exports (EQ/EPA, 1972).

Another set of estimates based on econometric analysis has focused specifically on the balance of payments effects of pollution control, starting with an estimate that environmental control costs would raise US costs and prices of tradeable products by about 3.6 per cent. Whether the US applies environment measures unilaterally, or whether these are paralleled by measures taken simultaneously in major competitor countries, these estimates conclude that there will be a net positive change in the US level of income of less than 3 per cent – small enough to be considered 'noise' in the econometric model used. The US balance of trade suffers, but only by about 3 per cent of 1968 exports in the unilateral case, and actually benefits by slightly less than one per cent in the multilateral case.[4]

Neither of these studies can be considered definitive, of course, and both contain large margins for error – both because of the notoriously poor quality of the data on environmental-control costs and because of even greater uncertainty regarding pollution costs that will be facing the major US competitors abroad. Nor do they identify those industries or sectors likely to feel a more substantial international competitive effect, or those for which the competitive impact is likely to be negligible.

Empirical investigation using input-output analysis shows quite clearly that the competitive effects on trade of environmental controls is highly industry-specific (Mutti and Richardson, forthcoming). The reasons for this are:

1. The wide differences that exist in the case of individual products in direct pollution-control costs per unit of output that could be passed forward onto product price – differences that are even more pronounced when indirect pollution control costs (via inputs) are included;
2. Differences in corporate spending on pollution control at the company level, depending in part on locational patterns of manufacturing facilities;
3. Differences in the degree to which international trade and competitive factors are important at the product level.

Frequently the pass-through effect of pollution control costs is in fact relatively small. In one study, of 28 US firms that reported such cost increases in 1971, only 8 reported any price increases, and another 6 reported the reason as being foreign and domestic competition (Henry, 1974, p.22). For example, in the non-ferrous metals industry a 20 per cent cost increase during that year occurred together with a 10 per cent price decrease resulting from international supply and demand shifts. In such industries the pass-through effect may thus be an intermediate or long-term phenomenon. In others, pollution control costs may principally affect profitability and hence be confined entirely to long-term supply effects.

Hence, whether or not fixed pollution control costs eventually affect product prices depends on competitive structure and pricing behaviour at the industry level. The US Council on Environmental Quality (CEQ) has proposed that the induced cost-shifts will be nominal for most industries when viewed in relation to the value of total shipments. CEQ also points to wide inter-industry differences in pollution control costs and capital allocations, as well as discontinuities over time in the growth of expenditures for this purpose – actually showing *declines* in petroleum, chemicals, paper and certain other non-durable goods industries between 1973 and 1974. Also to be considered, as noted above, are the indirect effects of pollution control costs on inputs into internationally traded products – what happens to output when these indirect effects are taken into account?

To summarise, if we assume that environmental policy is consistently aimed at passing the costs of pollution abatement forward onto product prices, the nature of the effects on international trade and competition are relatively straightforward. Abatement of process pollution will tend to raise production costs of final goods. Anti-pollution capital equipment will raise *fixed* costs to the firm. Efforts to combat pollution may also depend on production levels (e.g., operating costs) and will therefore raise *variable* costs to the firm as well. In a highly competitive market, environment-induced cost increases have relatively little effect on product prices but lead to relatively large cuts in output levels. In less competitive markets these costs will largely be rolled over onto product prices. In either case, if domestic producers face higher pollution control costs than competitive foreign suppliers, they will eventually lose in international competitiveness. At the national level, this will generate increased volumes of imports and reduced export volumes.

The principal policy issue here is the possibility of compensatory import charges sought by domestic suppliers in countries adopting the most stringent pollution controls, as a shield against competitive suppliers in foreign countries having less rigorous environmental requirements – e.g. as specified in the US Federal Water Pollution Control Act.

Since pollution control related to productive processes is a public-policy decision that influences production costs, it bears some relation to effective rates of protection (ERP) (Grubel, 1971). As in the case of occupational health and safety standards, the resultant cost increases will tend to erode effective protection enjoyed by domestic producers whether they are applied to finished goods or to intermediate and raw-materials inputs. One study has attempted estimates of the size of this erosion using United States data. Overall, pollution control costs were estimated to be 4 per cent of value added for all industry, while composite effective rate of tariff and non-tariff protection has an estimated overall value of 15 per cent, such that relative pollution control costs are about 27 per cent of the ERP (Walter, 1974). This seems to indicate that environmental control costs are not at all trivial when viewed in relation to the degree of value-added shielding afforded by national protective structures. It also illustrates the incentive that import-competing suppliers may have to secure a rise in the effective rate of protection in order to offset its prior erosion through pollution control costs that may be passed onto product prices. Again, to the extent the polluter pays principle is not followed, the erosion of effective protection will be correspondingly less.

Whereas there is considerable uncertainty as to the trade and competitive effects of environmental policy in the short term, there is even greater uncertainty about the long-range effects of environmental management on international comparative advantage. Restoration and maintenance of environmental quality represents production of a social good (or alleviation of a social 'bad') in which various proportions of available resources will be absorbed, presumably with maximum efficiency under given technological conditions. If pollution control activities are by their nature capital-intensive, which may well be true, then capital-abundant nations will tend to emerge with an international comparative advantage in something that cannot itself be traded, but that can be embodied in traded products, and thereby influence their long-term competitive position in the international marketplace. At the same time, those countries having within their borders environmental resources of substantial assimilative capacity will eventually develop a comparative advantage in the production and trade of goods and services whose supply characteristics tend to be pollutive in nature (Siebert, 1976).

Finally, it is important to include the role of multinational corporations (MNCs) in assessing the trade effects of environmental policy. A multinational company with many plants in many countries has the option of serving a particular market from several production points. If one of these points is located in a country whose environmental policy is generating significant incremental costs or supply blockages in manufacturing or in the procurement of inputs, it may indeed be possible to shift production of the affected item to a plant in another country, and subsequently serve markets from the new location – perhaps replacing the lost output in the old location with products that are less environmentally costly to manufacture. In the process, the patterns of international trade and production will tend to shift.

Given that MNCs today control a large and growing share of world trade, such inter-country production shifts may in fact occur to a significant degree *within individual multinationals*, rather than on an 'arm's length' basis involving competitive market-related gains and losses of purely national firms. Whether the decisions are internal to the firm or external in the competitive marketplace, however, the effects on trade, production and employment as viewed from a purely national perspective may well be much the same.

To the extent that they in fact react to environmental policy measures, therefore, it would appear that the existence of multinational firms may increase the *sensitivity* of the international economy to differential environmental conditions, policies, and instruments. They increase the speed of adjustment with respect to trade flows, and may well enhance their own competitive positions in the process. MNCs make within the firm some of the decisions about shifting trade flows that otherwise would be made by market forces altering the international competitiveness of individual suppliers. And MNCs may be able to draw on their internal technology bases and corporate information systems to secure advantages with respect to international markets for pollution control equipment and technology. This may eventually lead to a strong competitive export position for MNC home countries if they are indeed in the vanguard of environmental policy.

Trade and competitive effects: product pollution

Another link between environmental policy and international trade is through the pollutive characteristics of traded products themselves. The production of raw materials recovery process of traded goods may itself be deemed extraordinarily damaging to the exporting country's environment – or the environment of other countries via transfrontier pollution – so that trade restrictions may be called for. Or the traded products may be environmentally damaging in ordinary use. Or they may damage environmental quality when they are discarded at the end of their useful lives.

Traded products may cause damage to the *ambient* level of environmental quality either directly, or indirectly when combined with other elements. Their presence in the natural environment may damage the health of human beings or other life forms. Products containing damaging agents may also be subject to restrictions on environmental grounds even though the products themselves are entirely harmless, particularly food products. Because so many environmental factors affect health, it makes little sense to try to separate environmental restrictions from a subset of the more common health and sanitary restrictions.

For example, a major source of environmental problems associated with internationally traded finished and intermediate products involves the presence in the human environment of cancer-causing agents. Exposure to vinyl chloride and certain synthetic fuels are alleged to cause various types of cancer. Polychlorinated biphenyls have been linked to liver cancer, as well as skin infections, nausea, dizziness, asthmatic bronchitis, and fungus infections. Fluorocarbons – commonly used as refrigerants and aerosol propellants – are alleged to cause ozone-depletion in the atmosphere, thus increasing ultra-violet radiation and leading to increased skin cancer in humans. Indeed, recent studies have blamed environmental elements for up to 80 per cent of cancers in man. Carcinogenic substances may interact with one another in complex ways to cause a cancer that a single substance alone could not produce. While total avoidance of environmental carcinogens is clearly impossible in a complex society, the identification of particularly dangerous substances is increasing and the tradeoffs involved are becoming more clear. As a result, environmental policies in growing numbers are being applied to actual or suspected carcinogens in the environment.

Another example concerns agricultural pesticides. In late 1975 the Environmental Protection Agency ordered the suspension of the sale of two widely-used agricultural pesticides, chlordane and heptachlor, employed in certain farm crops, including corn and tobacco, to control soil insects. Both chemicals are suspected of causing cancer in humans, according to laboratory tests on animals. Both are chlorinated hydrocarbons similar to DDT (banned in the United States in 1972) and are highly useful in pest control. 21 million pounds of chlordane and heptachlor were produced in the United States in 1975, and 38 million pounds were planned for 1976 prior to the ban. This action will clearly put pressure on residues of the same or similar pesticides on or in imported agricultural products.

Other products involved include automobiles and other motor vehicles, in terms of emissions of carbon monoxide, particulates, sulphur dioxide and nitrogen oxide, all of which can have respiratory consequences. Mineral fuels in general are of considerable concern, particularly in terms of sulphur content, whether used in mobile or stationary combustion. Tetraethyl lead and other anti-knock compounds are damaging to health when ambient in the environment, and are subject to growing restriction in a number of countries. Asbestos can cause severe respiratory problems when fibres are present in the environment. Heavy metals and other inorganic substances, including compounds of mercury and cyanide, are widely used in manufacturing and are frequently discharged in quantity into the environment where, due to their non-biodegradability, they may have extremely long residence times. Effects on human health can include brain damage and genetic mutation. Radioactive substances and explosives can likewise have direct effects on human health if they are present in the environment. Others, like mercury, may have indirect effects in that their presence in the environment in minimal amounts may lead to a concentration in the food chain to possibly lethal doses at the point of ingestion by human beings.

There are a number of other internationally traded products that might be termed environmentally 'sensitive' yet do not pose an immediate threat to human health. This includes detergents that may pollute waterways and kill plants and animals. It also includes fertilizers which, through agricultural runoffs into bodies of water, may stimulate excessive growth of algae – eventually leading to oxygen depletion and extinction of other forms of life in lakes or rivers. Moreover, various kinds of packaging have been deemed environmentally damaging because of excessive use of materials or non-biodegradability, causing environmental damage associated with depletion of natural resources as well as accumulations of solid wastes and litter. Furs, skins, leather, bones and ivory from endangered species of animals, as well as products of the whaling industry, have also been subject to various sanctions in international trade for environmental reasons.[5]

Lastly, there are a number of products that affect the environment in less tangible and sometimes temporary ways. One example is noise pollution by aircraft and motor vehicles. Another is the possibility that widespread use of supersonic transports may alter the global ozone layer or, through formation of ice crystals, alter the earth's temperature (d'Arge, 1976). Here restrictions on product *use* may seriously affect production and trade in the items in question.

Trade barriers imposed for reasons of environmental policy and affecting pollutive products may take a variety of forms, which may be listed as follows:

1. Prohibitions. The most drastic environment-related measure may ban a particular product altogether. In such a case, the manufacture, storage, sale, use and disposal of the product may be prohibited, and all imports may be embargoed. The restriction is likely to be completely non-discriminatory, both between domestic and foreign suppliers and among sources of imports. Market-access is effectively closed off.
2. Global embargoes. In certain cases the manufacture of an environmentally sensitive product may have to be strictly controlled, which in turn may be possible only domestically. Hence imports from all sources are prohibited.
3. Selective embargoes. Products originating from certain sources of supply may be known to be environmentally unacceptable to importing countries. An example is the high mercury content of Ethiopian salt exports, which is considered excessive by Japan. As a result, imports from such sources may be banned while they continue from (or are diverted to) other supplier countries (UNCTAD, 1976).
4. Use restrictions. Certain environmentally damaging products, such as phosphatic detergents, may be restricted in use by individual countries or regions. As a result, demand tends to decline for both domestic and foreign suppliers.
5. Technical standards. The use of some products may be permitted at the national or sub-national level only if they meet certain environmental standards. Sulphurous fossil fuels and motor vehicles are the most prominent examples. Suppliers, whether domestic or foreign, must know precisely what these standards are and must meet or surpass them in order to secure market access.
6. Disposal standards. A number of products are environmentally damaging when they are disposed of, after their useful life ends. Packaging materials are a good example. Again suppliers must meet these standards or, if consumers are directly affected, may experience a downward shift in demand.
7. Export restrictions. Domestic environmental measures may restrict exports under certain conditions. Extraction of renewable or non-renewable resources may be environmentally damaging, so that access to foreign markets is deemed counterproductive in so far as it increases production levels. Products of endangered species may be subject to more or less strict export controls. Recycled waste products such as scrap metal may be subject to export restrictions in times of domestic scarcity. Such measures may deny foreign producers access to needed intermediate products and raw materials.

A recent survey by UNCTAD appears to indicate that trade barriers imposed for environmental reasons represent a potential problem that has not yet come to full fruition (UNCTAD, 1976). But environmental factors are beginning to affect market-access for a wide and growing array of internationally traded manufactures and semi-manufactures. For the developing countries it is still limited to food products, in large part. For the developed countries it is beginning to spread from automobiles into a variety of industrial products as well.

For suppliers involved in international markets, environment-related restrictions imposed on traded products may thus prohibit or restrict their use, or mandate significant design changes intended to make them environmentally acceptable. For suppliers, this may mean narrowed markets, increased research and development and production costs, increased product prices, shorter production runs, and reduced sales. As noted, from the standpoint of international trade such controls generally affect all products sold in a given market, whether imported or produced domestically, so that there is no discriminatory intent aimed at foreign suppliers. On the other hand, export suppliers may have to service multiple markets that are subject to multiple sets of environmental controls, so that scale economies may have to be sacrificed in comparison with import-competing suppliers in these target markets. The issue is often difficult to handle, because many products combine both positive and negative environmental characteristics. And whereas the positive may dominate under conditions found in one country, the negative may dominate in another.

Again, adaptation to environmental policies affecting pollutive products depends, in part, on the response-patterns of multinational corporations. Assuming that such measures are imposed in a non-discriminatory manner, product-adaptation to serve the market falls on both imports and home-produced goods equally, with incremental production costs possibly falling disproportionately on suppliers of imports. Yet multinationals may be able to avoid the incremental costs associated with short production runs and the like by concentrating output for particular markets with common pollution control requirements at single production points within the firm's global logistical network. Hence, while competitive geographic shifts in production patterns might not normally be associated with product-pollution controls, such shifts may indeed occur if the strategic response to these controls is internal to the planning structure of the multinational enterprise.

The same kind of pattern may develop with respect to intermediates and raw materials. Multinational firms with multiple sources of parts and components may be able to obtain these selectively, in order to ensure the required environmental characteristics in the final products. The vertically-integrated MNC may have within the firm multiple sources of supply, and in the face of environmental controls can shift to other sources producing substitute products while shipping output from the displaced source to less environmentally sensitive markets. Similarly, when environmental controls impair productive capacity at any one plant, it may be possible for the MNC to shift to other sources without disrupting the firm's overall operations.

The extraordinary technological capabilities of multinationals may also give them a competitive edge in complying with *changes* in pollution control requirements that bear on internationally-traded products. The MNC may be able to adapt more rapidly and more efficiently to such requirements, and hence their trade and competitive effects on the firm may be ameliorated or even reversed. That is, the competitive position of the MNC may indeed be *strengthened* as a result of its relatively effective compliance with product-pollution controls that cannot be matched by its non-multinational competitors (Gladwin, 1977).

Also of interest is the point that international differences in the pace of environmental policy give rise to the development of markets and trade in pollution-control hardware and technology. Countries that are in the lead in applying environmental measures tend to generate an internal demand for standard kinds of plant and equipment to deal with the problem, as well as innovative approaches for achieving the goal (Walter, no date). This, in turn, is likely to stimulate firms to meet the market demand thus created. When the technology and supply capability is in place for the domestic pollution-control market, it can easily serve as the base for serving export markets once the relevant pollution-control demand is created abroad. Hence the 'product-cycle' of pollution-control equipment, to the extent that it embodies innovative technology, may favour those countries leading in the area of environmental policy. The competitive edge created in this way for lead-countries may serve partially to offset the competitive disadvantage that some of these same countries may suffer with respect to trade in conventional pollution-intensive goods and services. It may apply not only to 'pure' environment-control and monitoring equipment, but also to standard types of machinery that are more environmentally efficient and so may be used to replace older and more pollutive equipment.

Lastly, environmental policy may also give rise to substantial international trade in secondary materials. The amount of recycling of in-plant and 'old' scrap metals, paper, textiles, wood and other materials is highly dependent on prevailing market prices of both virgin and secondary resources. The existence of international scrap markets may serve to keep domestic prices higher than otherwise would be the case, and thereby help to promote secondary materials recovery and reuse – and simultaneously discourage injecting the waste into the environment. At the same time, strict pollution control policies will tend to raise the effective cost of waste disposal, and thereby once again help to encourage recycling. There is no reason why the international economy cannot promote global efficiency in the reuse of society's wastes, just as it promotes global efficiency in the production process itself. But this requires essentially free trade in the waste-material sector (Grace, Turner and Walter, 1976).

Environmental policy and plant location

Aside from its influence on trade via differential production and product costs, environmental policy may influence international economic relations by affecting plant-siting decisions by multinational firms. Conceptually, a distinct cause-and-effect relationship between pollution control and trade via multinational corporate industrial siting decisions is just as relevant (if not more so) for long-range evaluation of its international competitive impact as are the more direct trade-effects via relative prices.

It may well be that, in a particular case, local environmental conditions (or anticipated future conditions) will preclude corporate management from going ahead with plant expansion or with the establishment of a new facility. Management then has the choice of moving elsewhere in the US, moving abroad, or abandoning the plan and searching for other investment outlets.

Given interregional and/or international differences in environmental policies and priorities, decisions related to the location of new industrial facilities or the expansion and relocation of existing ones will tend to be influenced accordingly. The anticipated effects depend on

1. The importance of pollution control as a cost element,
2. The extent of international differences in environmental standards and their application,
3. Differences in the availability and quality of labour, public services, transport costs and proximity to markets,
4. The nature of a given firm's internal logistical network, the character of its products and optimal marketing strategy, and its rate of growth in required productive capacity.

Any such impact on the international economy will tend to be felt initially within specific industries, and progressively spread to broader ranges of economic activity. Firms that find preferred sites excluded for environmental reasons have available to them alternatives that encompass both domestic and foreign locations, and there are bound to be certain international locational spillovers attributable to local, regional or national environmental control. Processing may be done closer to the sources of primary materials supplies, particularly in developing countries, resulting in a rise in the average value of the latter's exports. Environmental considerations may, for example, contribute to the development of refineries and petrochemical plants in petroleum-exporting countries, with higher-value 'product' rather than crude oil and gas entering trade channels.

There is as yet only limited evidence of cross-border locational shifting by MNCs in response to international differences in environmental standards and concomitant pollution-control costs. Certain copper smelters, petroleum refineries, asbestos plants and ferroalloy plants have reportedly been constructed abroad rather than in the US for environmental reasons. A number of petrochemical complexes and chemical plants originally slated for West Germany and the Netherlands have evidently been resited in Belgium, France and Spain. And some recent Japanese pollution-intensive investments have reportedly been channelled to developing nations in south-east Asia and Latin America. But the evidence, viewed on a world scale, certainly does not suggest massive environment-induced locational shifting thus far (Gladwin and Walter, 1976).

The locational spillovers that have occurred seem to stem from a blockage or exclusion of preferred sites in various developed nations, and not from an opportunistic search for low-cost environmental locations on the part of MNCs. A number of jurisdictions in the United States, Japan, the Netherlands, West Germany, Australia, the United Kingdom, and elsewhere have begun to discourage the location of new pollutive industry, following an implicit cost-benefit assessment at the political level that the negative environmental consequences outweigh the expected economic gains. MNCs that have met locational impasses in these regions have simply been forced to resort to alternative domestic and foreign locations where the 'ecology-growth trade-offs' are different.

Any substantial international relocation of industrial production for environmental reasons would, of course, have far-reaching economic consequences. On the financial side, it would put pressure on source-countries' international payments. But of greater concern are the prospective employment and income consequences resulting from environment-induced locational shifts. Migration of industrial capacity will tend to have a depressing effect on both, but it will not be symmetrical with respect to regional or industrial structure. Some industries and regions will tend to be much more heavily affected than others, thereby increasing the seriousness of the problem from the point of view of domestic public policy.

The available evidence noted shows a certain amount of geographic mobility of production in cases where major projects are blocked for environmental reasons. This is particularly true of European and Japanese MNCs, where the United States is often mentioned as a *favourable* location alternative. Environmental restrictions on transportation facilities are also mentioned as a factor in determining plant location. Various explanations have been offered as to why significant locational shifting has not yet occurred (Gladwin and Walter, 1976). It nevertheless seems clear that the ultimate international economic impacts of environmental policy may in large measure make themselves felt through locational variables, as industrial site-selection and environmental conflict assume greater importance in the future.

Summary and conclusions

This paper has surveyed the various interactions between international trade and environmental policy. It has identified the differential impact of pollution control on production costs and their reflection in product prices, and hence international competitiveness. Even after the environmental costs associated with intermediate and raw material inputs have been considered, the effects on most industries may not be very significant in relation to other factors determining international competitiveness. On the other hand, if some countries depart significantly from the polluter pays principle while others adhere to it, there may be some competitive disruptions in certain sectors. These, in turn, may give rise to demands for protection in order to compensate for differential costs-effects of environmental policy.

Whereas the majority of these effects may work through shifting market relations, a significant proportion may also work through internal decisions of multinational firms as to where they buy and where they sell. Accurate analysis of the issues must therefore combine both market and managerial response patterns.

The second area of concern focuses on pollutive products. Here national environmental policy may restrict or ban particular kinds of internationally traded goods, and/or set specific environmental performance levels. These must be coupled with trade restrictions in order to assure that imports meet the same standards required of domestically-produced products. At the same time, such restrictions may lead to protectionist non-tariff barriers that discriminate against imported sources of supply. This area probably constitutes the most serious potential threat to the liberal trade emanating from environmental policy. At the same time, environmental factors may give rise to a variety of new products in the pollution-control and monitoring area, which may turn out to be important offsets for firms otherwise subject to negative competitive influences from this source.

Perhaps the most important environment-induced source of international economic disruption may ultimately be delays and blockages of new plants resulting from environmental conflict. These may trigger interregional and international shifts in plant location, which in turn may have important trade and competitive implications.

One issue that may further complicate the trade aspects of environmental policy is transfrontier pollution (TFP). This involves the direct impact of pollutive activities, carried out in one country, on one or more other countries. In one-way TFP (as in a river system) the polluting country (upstream) imposes significant environmental costs on the victim country (downstream) without itself being subject to such costs. In two-way TFP (as in a lake bordered by several countries) the victim and the polluter country are both subject to environmental damage, and so the incentive to remove pollution is greater in two-way than in one-way pollutiion. Essentially, the TFP issue boils down to a question of rights – the right to pollute versus the right to a clean environment – although the incentive structures are not symmetrical.

Transfrontier pollution may give rise to retaliation on the part of the victim(s), given that the international legal machinery for the adjudication of TFP disputes does not yet exist. Such retaliation may take the form of trade barriers. For example, threatened extinction of endangered species is seen widely as a TFP issue, and has given rise to import restrictions in many countries of certain furs and skins as well as whale products. Other products that are alleged to cause TFP – or whose production is alleged to cause it – may also face restrictions in the future. An example is the supersonic transport aircraft. More narrowly, victim countries may impose trade sanctions on polluter countries in order to raise the cost to the latter of engaging in TFP.

Reducing the potential trade policy impact of environmental measures can conceivably take several forms. Product-related environmental measures imposed on all suppliers, whether foreign or domestic, are essentially non-negotiable. Hence they are very similar to health and safety standards, and prospects for negotiated removal are slim. Harmonisation of environmental standards among importing countries is essentially the only way the problem can be alleviated and – with possible exceptions such as the European Communities – this will be very slow in coming. Movements towards such harmonisation, particularly among the developed countries, should probably be encouraged. If, on the other hand, discrimination exists between domestic and foreign suppliers, or among foreign suppliers, the issue of negotiation becomes relevant and may be handled in ways similar to non-tariff barriers.

Attempts to establish increased tariffs on other trade barriers as a way to offset domestic production cost increases attributable to pollution control costs cannot be justified on economic grounds, since it would compound the misallocations of resources. If the fact that countries differ in environmental preferences or assimilative capacities leads to an alteration of trade and production patterns, this induces a more efficient use of environmental resources world-wide. Environment is one of the determinants of international comparative advantage, and should be allowed to operate accordingly. Hence compensatory import restrictions should be subject to the standard complaint and negotiation procedures existing in international forums and, if necessary, may require retaliatory commercial policy action or appropriate compensation.

Notes

1 For a more detailed analysis see Walter (1975).
2 For a survey see UNCTAD (1976).
3 See, for example, Pearce (1976).
4 Estimates by The Conference Board, New York, 1976 (unpublished).
5 For example, the US imposes quantitative import controls of products from endangered species of birds and animals.

References

Ayres, R.U. and **Kneese, Allen V.** (1969) 'Production, consumption and externality', *American Economic Review*, June.
Barde, Jean-Philippe (1976) 'National and international policy alternatives for environmental control and their economic implications' in Walter (1976), pp. 137ff.
CEQ/EPA/Department of Commerce (1972) *The Economic Impact of Pollution Control*, US Government Printing Office, Washington DC.
d'Arge, Ralph C. (1976) 'Transfrontier pollution, some issues on regulation' in Walter (1976), p. 257ff.
Gladwin, Thomas N. (1977) *Environment, Planning and the Multinational Corporation*, JAI Press, Greenwich, Connecticut.
Gladwin, Thomas N. and **Walter, Ingo** (1976) 'Multinational enterprise, social responsiveness and pollution control', *Journal of International Business Studies*, Fall.
Gladwin, Thomas N. and **Walter, Ingo** (1977) 'Multinational enterprise and the natural environment: diversity in challenge and response' in Kapoor and Grub (1977).

Grace, Richard, Turner, R. Kerry and **Walter, Ingo** (1976) 'Environment, international trade and waste materials', paper presented at the Southern Economic Association annual meetings, 18 Nov. 1976.

Grubel, H.G. (1971) 'Effective tariff protection: a non-specialist guide to the theory, policy implications and controversies' in Grubel and Johnson (1971).

Grubel, H.G. and **Johnson, H.G.** (eds) (1971) *Effective Tariff Protection*, General Agreement on Tariffs and Trade, Geneva.

Henry, H.W. (1974) *Pollution Control: Corporate Responses*, Amacom, New York, p. 22 displacement from environmental control: the quantitative gains from

Kapoor, A. and **Grub, P.** (eds) (1977) *The Multinational Enterprise in Transition*, Darwin Press, Princeton NJ.

Mutti, John H. and **Richardson, J. David** (forthcoming) 'International competitive displacement from environmental control: the quantitive gains from methodological refinement', *Journal of Environmental Economics and Management*.

OECD (1972) *Problems of Environmental Economics*, Paris.

Pearce, David W. (1976) *Environmental Economics*, Longman, London.

Siebert, Horst (1976) 'Environmental control, economic structure and international trade' in Walter (1976),

UNCTAD (1976) 'Implications of environment policies for the trade prospects of developing countries: analysis based on an UNCTAD questionnaire', Mimeo., Geneva.

Walter, Ingo (1973) 'The pollution content of American trade', *Western Economic Journal*, March.

Walter, Ingo (1974) 'Pollution and protection', *Weltwirtschaftliches Archiv*, March.

Walter, Ingo (1975) *International Economics of Pollution*, Macmillan, London, and Halsteed-Wiley, New York.

Walter, Ingo (ed.) (1976) *Studies in International Environmental Economics*, Wiley, New York.

Walter, Ingo (no date) 'Economic response of multinational companies to environmental policy: report to the US Department of Commerce', NYU/GBA working paper No. 76-88 mimeo.

Chapter 12

The role of international organisations in the control of transfrontier natural resources and environmental issues *M. Potier*

Environmental concerns emerged as a factor of government policy late in the 1960s, when some of the world's more industrialised countries, under growing pressure from public opinion, decided to set up institutions especially responsible for such matters.

While this is not to say that governments had previously framed policy with no regard for the environment, the creation of fully-fledged Ministries, Departments and Agencies to monitor and protect it was clearly a major step forward.

The need for international co-operation then soon became evident largely for two reasons:

1. In the first place, these new concerns caught many governments unprepared, since they usually lacked accurate scientific data, reliable pollution monitoring and measuring facilities, sophisticated methods for analysing processes, a suitable administrative organisation and even any consistent rationale or doctrine;
2. Governments had to reckon with the fact that many environmental questions are intrinsically international: some because they effect the planet as a whole (major ecological problems such as the balance between resources and population, climatic changes, genetic hazards, etc.); others merely because pollution can cross international frontiers (long-range transport of pollutants in the atmosphere and watercourses), or affect a shared environmental medium (such as a lake).

All this has given the international organisations a unique role over the last few years, which is the rapid dissemination of information and experience to better guide the policy-maker, and the promotion of international codes of conduct designed to prevent or help settle transfrontier pollution disputes.

The part played by any one international organisation has of course been in line with its own particular function, which is usually determined by its particular membership (industrialised eastern or western countries, developing countries). Thus such specialised organisations as World Health Organisation (WHO), Food and Agricultural Organisation (FAO) and United Nations Environment Programme (UNEP), consisting of countries belonging to the United Nations, tend to be more concerned with problems possessing a world dimension, and provide a forum for discussion between the eastern and western industrialised countries and the developing countries. The Economic Commission for Europe, a United Nations regional body located in Geneva, encourages the exchange of information and experience between West and East. Within the narrower confines of the Europe of the Nine, the European Community is trying to put together a community-wide environmental policy. In the OECD are represented the governments of the western industrialised countries, and they are clearly a long way from any common approach to environmental problems, in view of their different socio-economic and geographical structure, varying environmental patterns, national traditions and degrees of economic maturity. Common to them all, however, is the market economy system and it is in these countries, which account for more than two-thirds of world trade, that living standards are highest. Environmental questions therefore constituted an ideal field for co-operation within the OECD, whose Council decided in 1970 to set up a new plenary body, the Environment Committee, where government representatives could meet to discuss common problems, examine possible solutions, and where appropriate formulate recommendations and guidelines for national policy. Since environmental policies can have wide-ranging implications, representatives are drawn not just from Environment and Health Ministries but also from those dealing with Foreign Affairs, Finance and Industry. Naturally enough, as an organisation for economic co-operation, it focuses particularly on the various needed trade-offs between the economic and environmental aims of its Member countries. The mandate of the Environment Committee, renewed by the OECD Council in 1975 for a period of five years, provides that:

The Environment Committee will be responsible for:

> examining on a co-operative basis common problems related to the protection and the improvement of the natural and urban environment with a view to proposing acceptable solutions to them, taking into account all relevant factors, in particular economic and energy considerations;
> reviewing and consulting on actions taken or proposed by Member countries in the environment field and assessing the results of these actions;
> providing Member Governments with policy options or guidelines to prevent or minimise conflicts that could arise between Member countries in the use of shared environmental resources or as the result of national environmental policies; the Committee may organise as appropriate, and with the agreement of the countries concerned, consultations to that effect.
> encouraging wherever appropriate, the harmonisation of environmental policies among the Member countries.

The work of the Committee is organised around well-defined projects spread on average over two years; these usually lead to a recapitulative report with conclusions for consideration by governments in designing and implementing their environmental policies. These reports sometimes lead to formal acts of the OECD Council – Decisions or, more usually, Recommendations.[1]

The Committee is assisted by a number of specialist subsidiary bodies composed of government officials concerned with more specific aspects of environmental policy, such as air, water, chemicals, recycling, urban development, noise, energy. Other specialist groups study the economic implications of environmental policy and the legal consequences of pollution which transcends frontiers.

The Committee also relies on the work of its multidisciplinary international Secretariat, backed by administrative and technical staff belonging to the Environment Directorate.

This Committee has achieved concrete results in a number of areas since its inception. A tentative assessment of OECD action during the past eight years will show that it has proceeded on three main fronts:
1. Promoting the dissemination of knowledge regarding the environment through exchanges of information and comparisons of experience;
2. Promoting the harmonisation of environmental policies;
3. Furthering the concept of international solidarity in order to solve transfrontier pollution problems.

The rôle of the OECD in disseminating knowledge and comparing experience regarding the environment

It would be beyond the scope of this section to give any exhaustive account of the OECD's action over the years in disseminating information or comparing experience. Only a few of the topics which the Organisation has been working on[2] will therefore be mentioned as adequately illustrating the environmental rôle.

Dissemination of knowledge
A particularly broad area had to be covered owing to the diverse nature of activities having to do with environmental considerations. The major emphasis has however been on identifying the various forms of disamenity and pollution, their causes, effects on health and well-being and especially on ways of eliminating them or reducing their impact and ascertaining the cost of such remedies. On the other hand little work has been done on nature conservation, the preservation of species or the creation of nature parks, i.e. on the positive side of environmental policy, not because these are regarded as secondary or calling for international co-operation, but because they are covered by other international organisations (such as the Council of Europe). It was thus felt important to avoid any scattering of effort and overlapping. The OECD, through research by its Secretariat and with the assistance of specialised expert groups set up and hence promoted exchanges of information in such fields as air and water pollution, chemical and noise pollution, meanwhile investigating a number of what are probably the most heavily polluting industries (pulp and paper, energy, iron and steel, petrochemicals, aluminium, fertilisers, automobile and aircraft industries).

Information exchanged has also mainly dealt with the emission of pollutants generated by these industries, the kinds of technology which might help to reduce them and the cost involved.

By way of illustration the results of a recently completed five-year programme to measure the propagation of air pollution over Europe will be described in greater detail. This exercise in international co-operation (OECDj) by eleven OECD Member countries (Austria, Belgium, Denmark, Finland, France, Germany, Netherlands, Norway, Sweden, Switzerland and the United Kingdom), making use of seventy ground sampling stations and a large number of aircraft measurements, yielded findings of obvious value to programmes for sulphur pollution control at national as well as on an international scale.

High concentrations of sulphur in the atmosphere are a potential hazard to human health. In some instances they are suspected of causing bronchitis and other respiratory disorders, shown by epidemiological research to reduce the life expectancy of individuals exposed. At lower concentrations, such as those dealt with by the OECD study, airborne sulphur is shown to cause acid rain and snow whose effects have included the destruction of fish in various watercourses and lakes, particularly in Norway and Sweden and perhaps also in Scotland.

The programme has confirmed what there had already been some reason to suspect, that sulphur can be transported over several hundreds or even thousands of kilometres by air currents; the Scandinavian countries moreover prove to be not the only ones affected. About half of the sulphur received in Finland, Norway and Sweden comes from sources in other countries, and this is also true for Belgium, the Netherlands and Switzerland, while the figure for Austria is two-thirds. In contrast, less than one-tenth of the sulphur deposited in the United Kingdom arrives from outside. In the other countries examined (Denmark, France and Germany) sulphur from abroad accounted for some one-third of the total deposited.

The findings of this programme are of particular interest because they show that apart from the United Kingdom, no European country can independently control the level of sulphur pollution affecting its territory. This exchange of information thus takes on a clear policy 'dimension' once international co-operation among the countries of Europe is recognised as a necessity. Hence the OECD's present development of a further programme to see what measures governments might take to bring sulphur pollutants under international control.

Owing to the special character of the Organisation considerable attention is naturally paid to the economic aspects of environmental policy. Many questions of implementation thus arise. On what scale should resources be committed to pollution control? What benefits are such programmes likely to yield? How will they affect GNP growth rates? Will they increase unemployment or push prices up? What will the effect be on individual regions or industries, or on the balance of payments? Answers to some of these questions are emerging from the work of the Group of Economic Experts and the Secretariat (OECDa,b,c,d,e,h), showing that value-added (direct and indirect) absorbed by pollution control programmes yearly averages some 1 per cent of GNP for most countries, except the United States and especially Japan, for which the figures are higher. Their work has also shown that the economic consequences of implementing pollution control programmes are apparently of little significance.

Comparisons of experience

The comparing of experience is another significant aspect of international co-operation. The Organisation has played a special rôle in recent years by serving as a forum for the industrialised countries of the Western world. On several occasions Member countries have compared their experience in particular environmental matters, thus pooling to their mutual advantage the hard lessons afforded by an analysis of successes and failures. The usual procedure, on the basis of an analytical framework jointly determined and prepared by the Organisation's Secretariat, is to have the countries concerned or the Secretariat draft case studies dealing with factual experience and designed to provide those responsible with data relevant to a number of basic issues, so that lessons can be drawn for the future from the past and present experience.

Such an approach has enabled the Secretariat, with the help of a few consultants, to make an assessment of economic instruments (pollution charges) used by OECD Member countries in implementing environmental policies (OECDi). This study indicates that pollution charges are still comparatively little used, except perhaps for water pollution, but that they are gradually being applied to new fields (noise, waste, etc.). Public acceptance of such instruments depends essentially on how simply the charges can be worked out, how the funds collected are redistributed, and how far the parties concerned share in implementing the charging systems.

A more recent decision has been to compare the experience of several Member countries with heavily industrialised hydrographic basins, hence marked by special arrangement difficulties. It was agreed that this analysis should be confined to the following important points:

1. Determining how standards for discharge into the aquatic environment are laid down, and how they are linked to quality standards in cases where such standards are decided by the authorities responsible for water policy;
2. Establishing how far discharge and quality standards are effectively applied by the different categories of polluter;
3. Ascertaining how far the public in the broader sense, and the various categories of water-user, participate in the water management decision-making process.

Another method of promoting the comparison of experience is for a country to submit its overall policy to a review in depth. On the one hand, the country concerned may have its attention drawn to problems eluding it in day-to-day management but which outside specialists can identify, and may thus reap the benefit of constructive criticism by the international community. On the other, countries participating in the review can also benefit from a detailed knowledge of some other country's policy whose methods it hopes to adapt wherever possible while avoiding its mistakes. This kind of review is of course only worthwhile if it is scrupulously carried out and provided that the strong and weak points of the country's policy are assessed as objectively as possible.

Since the establishment of the Environment Directorate of the OECD, environmental policy has been examined in two countries: Sweden in 1973 and, more recently, Japan in November 1976. In both cases, the Secretariat prepared an account of policy in the country concerned, and an Analysis Meeting was attended there by high-level environmental specialists from OECD countries.

The review of Japanese environmental policies was particularly instructive since it is the industrialised country which has suffered more heavily from pollution than any other, mainly because of its very rapid industrialisation and high population density.

The review concentrated on four main themes:

1. The setting of standards (for ambient air quality, product standards, emission standards);
2. The Japanese system for compensating pollution victims;
3. Environmental problems associated with new installations;
4. The cost and economic repercussions of environmental policies.

Noteworthy findings were that:

(a) Japan has managed to conduct its very ambitious environmental policy without jeopardising either the country's economic growth or the competitiveness of Japanese industry. The severe standards imposed on industry strongly stimulated technological innovation without unduly increasing costs;
(b) despite the remarkable progress achieved in controlling air, chemical and noise pollution, much remains to be done to improve the quality of life, i.e. the whole range of unmeasurable components of life such as aesthetics, the right to privacy and social relations.

It is a matter for some satisfaction that in the light of this review the Japanese Government has decided to set up a high-level working group to propose ways of improving the quality of life in Japan.

The rôle of the OECD in harmonising environmental policies

Because of the varying local situations, especially in regard to ecology and the OECD countries' wide spectrum of socio-cultural values, there may be some doubt as to how desirable or realistic it is to try to harmonise environmental policies, i.e. more closely align policy objectives if not achieve some degree of convergence in policy methods. Harmonisation should not of course be an end in itself, but it appears to be desirable whenever obstacles to trade among OECD Member countries threaten to interfere.

This point has been taken, and in 1972 the Member countries adopted a set of 'Guiding Principles concerning the International Economic Aspects of Environmental Policies', including the so-called Polluter-Pays Principle which aims at basing environmental policies on common economic principles. The desire to obviate trade barriers, combined with their concern for environmental protection, moreover led the OECD Member countries in 1977 to adopt a Recommendation establishing guidelines for anticipating the effects of chemicals on man and his environment. The Guidelines and Recommendation will be discussed in turn.

The effect of adopting the OECD Guidelines on the harmonisation of environmental policies in the industrialised countries

The OECD was one of the first international organisations to adopt a set of guidelines to promote the harmonisation of environmental policies. While the Polluter-Pays Principle is the best known, the rules dealing with the harmonisation of standards should not be overlooked.

Adoption of the Polluter-Pays Principle by the international community of western industrialised countries reflects their desire to adopt a common, economically effective guideline as the basis for pollution control, and avoid difficulties bound to affect international trade if different guidelines were adopted. This principle means that the cost of pollution prevention and control measures should be borne by the polluter and expresses a concern for efficiency (through the internalisation of external effects) in pollution control policies. It essentially implies that no subsidy should be granted to a polluter to finance the cost of pollution control or prevention measures. Clearly, then, in aiming at greater economic efficiency by adopting the Polluter-Pays Principle, the OECD Member countries also sought to reduce the potential differences for industry should polluters fail to allow for the costs of pollution prevention and control measures and to include them in production costs. While the idea is that the Polluter-Pays Principle may be implemented gradually, stage by stage, implying that subsidies may be granted on a transitional basis, a condition is that no significant distortion shall result for international trade and investment (see OECDf, pp. 29–30), since adoption of the principle at international level should do away with any obstacle to international trade.

To monitor the application of this principle, a procedure had been developed for notifying Member countries through the Secretariat regarding all forms of aid. According to the figures obtained under this procedure, aid granted by Member countries during 1973, 1974 and 1975 amounted to less than one-tenth of 1 per cent of GNP, and so far such aid is unlikely to have caused any significant trade distortions. Continued operation of the procedure will make it possible to check the validity of these conclusions for the future.

When they adopted the Polluter-Pays Principle the OECD Member countries also agreed on other rules dealing in particular with environmental standards. Since the level of environmental standards (emission and quality standards) adopted in a country depends on such factors as the capacity of the environment to assimilate pollution, the degree to which the particular country is industrialised, the population density, and how much priority its citizens attach to protecting the environment, any very high degree of international harmonisation for standards will clearly be difficult to attain in practice.

The Member countries of the Organisation have recognised that 'where valid reasons for differences (between standards) do not exist, Governments should seek harmonisation of environmental policies, for instance with respect to timing and the general scope of regulations for particular industries, to avoid the unjustified disruption of international trade patterns and of the international allocation of resources which may arise from diversity of national environmental standards'.[3]

In considering how far it might be desirable to harmonise the standards internationally, they made a sharp distinction between three categories: environmental quality standards, emission and process standards and product standards.

With regard to the first category, since such standards are means of attaining quality targets set for a particular environment, and hence must be established at geographical level, they will obviously vary from one country or region to another. However, such standards should preferably be harmonised by countries deciding to form some economic or political union reflecting their political determination to implement quality objectives, when the geographical areas concerned constitute an environment calling for joint protection or one that is subject to transfrontier pollution, or for the international control of persistent toxic substances.

For the second category, international harmonisation of emission and process standards would seem all the less desirable whenever these standards are adjusted according to the particular area and to standardise them would be economically wasteful. Only in frontier regions or in the case of persistent toxic substances, would harmonisation appear appropriate.

As regards the third category, when products affected by standards of composition, design or utilisation are traded internationally on a substantial scale and the standards laid down by some individual country threaten to create a serious non-tariff barrier to such trade, harmonisation would clearly be desirable.

OECD action to promote the international harmonisation of control over chemicals

In recent years Governments have become increasingly concerned about the dangers to man and the environment entailed by the production, use and elimination of proliferating chemical substances. The figures indicate that some 50,000 chemical products are currently in use, with several hundred new ones coming on to the market each year. This concern has recently been reflected by the adoption in many countries of new legislation generally supplementing existing laws and regulations applying to such specific fields as dangerous products and toxic substances, medicines and cosmetics.[4] Although not all these laws have identical aims, they all call for the systematic study and evaluation of the effects of a chemical before it is manufactured or used on a large scale, to determine what risk it may constitute for man or the environment. Once countries began to apply such laws, a danger was that their implementation procedures would not match, thus causing a waste of resources and at the same time preparing the ground for non-tariff barriers to trade.

Thus the Recommendation setting out guidelines and the steps needed for assessing the potential effect of chemicals on man and the environment was adopted by the OECD countries at the right time, because it defines a concerted approach. The Recommendation can be briefly described as proposing five guidelines. The first defines the types of product which should ordinarily be subjected to systematic assessment. The second describes the various stages in the procedure to be followed, the kinds of test required and the order in which they should be carried out. The third deals with appropriate administrative procedures for this type of systematic evaluation. The fourth establishes the principle that the substance should be accompanied, whatever its destination, by certain particulars concerning the uses to which it can be put and appropriate precautionary measures. The fifth deals with activities to be put in hand at a later date to detect any effects which may not have been discovered during the assessment itself.

Adoption of this Recommendation was a step forward in promoting the interchange of data on chemicals among countries, since all Governments will be requiring industrial companies, both domestic and foreign, to provide similar information. But further problems remain, one being the reliability of data exchanged. Work was therefore begun, as soon as the Recommendation had been adopted, to reach agreement on a number of methods for testing the potential environmental effects of chemicals which might be approved internationally, thus making it easier for one country to accept data provided by another.

The rôle of the OECD in promoting international solidarity and in settling transfrontier pollution problems

The fact that the environment is largely composed of air and water, which are mobile and capable of transporting pollution from one region to another, naturally adds an international dimension to environmental problems, since in some forms pollution can cross frontiers and cause damage in countries where it did not originate.

Furthermore, the environment often takes the form of a common or shared resource (seas, lakes), belonging to no one individual or nation but to several collectively. Governments therefore have to agree on how to use the common or shared resource and must jointly lay down rules for its protection.

In these circumstances disputes are apt to arise between Governments, and international action is clearly necessary to resolve or prevent them. This may take different forms, as by adhering to a multilateral convention or treaty[5] or negotiating a code of good behaviour in a multilateral framework.

In recent years the OECD has played a part by urging its members to accept a number of responsibilities and obligations as a minimum and to prohibit certain types of behaviour.

The Organisation's activities in this field are largely articulated around three principles – international solidarity; non-discrimination; information and consultation at government level.[6]

The principle of international solidarity

The principle of international solidarity is based on the idea of an equitable balance between countries' mutual rights and obligations. To give this principle concrete form the OECD Member countries are invited 'to define a concerted long-term policy for the protection and improvement of the environment in zones liable to be affected by transfrontier pollution'. Meanwhile 'countries should individually and jointly take all appropriate measures to prevent and control transfrontier pollution, and harmonise as far as possible their relevant policies'. Furthermore 'countries should endeavour to prevent any increase in transfrontier pollution, including that stemming from new or additional substances and activities, and to reduce, and as far as possible eliminate, any transfrontier pollution existing between them within time limits to be specified'. It will be apparent that these provisions prepare the ground for a code of good behaviour which can serve as a base of reference in future negotiations among the various Member countries.

The principle of non-discrimination

The concept underlying the principle of non-discrimination is that a polluter who may inflict damage in some other country should be subject to legal or statutory provisions no less severe than those which would apply for any equivalent damage caused in his own country under comparable conditions. This means, in particular, that if a country applies the Polluter-Pays Principle to polluters located within its jurisdiction it will do so regardless of whether the effect of the pollution occurs inside or outside its frontiers. In other words, the corollary of the principle of non-discrimination is the principle of equality of treatment, not only for the polluters but also for the victims. A victim suffering damage as a result of transfrontier pollution can hence expect to be treated at least as favourably in the polluting country as would be a victim of internal pollution under similar circumstances.

The recognition of a transfrontier victim's same rights as an internal victim of pollution under similar circumstances implies, procedurally, that in seeking relief, he has the same rights of standing in judicial or administrative proceedings as an internal pollution victim.

The principle of information and consultation at intergovernmental level

One way of facilitating the solution of transfrontier pollution problems at intergovernmental level is through acceptance by the countries concerned of the principle that they should provide one another with any relevant information and consult one another as necessary. According to the OECD Recommendation, governments should receive advance warning of any proposed activities which might threaten transfrontier pollution in a neighbouring country, and should be able to hold consultations with the government of that country. In the course of this consultation, the country liable to be affected could request the country of origin to apply the principle of non-discrimination. The consultation should usually enable an agreement to be reached, provided the negotiations are conducted in good faith.

The principles adopted in conjunction with a number of OECD Recommendations on transfrontier pollution do not constitute legally binding obligations on Member Governments. Nevertheless the fact that they have been adopted represents an important step towards a combined approach, since the concept they embody will provide the basis for further initiatives in the fields of international law and co-operation with respect to the environment.

Conclusion

It may be helpful to conclude this brief review of the OECD's environmental activities by first defining its rôle with respect to the full range of functions which presumably should be performed at international level, and then in terms of tasks as divided among the various international organisations.

The functions to be exercised at international level were classified, in the action plan adopted in 1972 at the Stockholm Conference, under three headings:
1. Assessment of the environment (evaluation and review, research, monitoring, information exchange);
2. Management of the environment (goal setting and planning, international consultation and agreements);
3. Supporting measures (education and training, public information, financing, technical co-operation).

Among these, the OECD has played a special rôle by promoting the exchange of information on national surveys and projects, by evaluating and reviewing environmental policies, with particular respect to their economic consequences and in the formulation of new rules proposed to Member countries for adoption in the form of Recommendations.

As has already been pointed out, while not binding, the Recommendations imply a moral commitment on the part of the adopting Member countries to carry them out. The rule-making action of the OECD is hence quite different from that of the European Communities, since the Council and the Commission of the European Communities can adopt regulations which are binding on Member States, but very similar to that of the great majority of international organisations, which are not empowered to adopt regulations which become binding upon their Member States. Usually observance of the Recommendations is monitored, although this must not be taken to mean, for instance, that international police patrols are deployed on the high seas to thwart polluters. This kind of approach might seem ideal from several points of view, but is still practically non-existent. It is much more usual for countries to undertake such monitoring activities themselves and report to the international organisation appointed for the purpose on how their authorities are implementing the Recommendations. It is in this way that the OECD can check on how the many Recommendations adopted under its auspices are being applied.

Tasks have been divided among a fairly large number of international organisations largely according to two criteria: the ordinary jurisdiction of an organisation and its geographical scope. Thus, a distinction is naturally made between organisations of a world-wide scope and those with a far more regional function.

The United Nations performs a leading world function, particularly through the United Nations Environment Programme established after the Stockholm Conference.

The functions of UNEP are fairly different from those of other international organisations, since its purposes are above all to provide leadership, planning and co-ordination. It has a special rôle in co-ordinating environmental activities undertaken by the specialised institutions of the United Nations: (Food and Agricultural Organisation (FAO), United Nations Educational, Scientific and Cultural Organisation (UNESCO), World Health Organisation (WHO), World Meteorological Organisation (WMO), Inter-governmental Maritime Consultative Organisation (IMCO), International Labour Organisation (ILO).)

The regional organisations are mainly European, or organisations with their centre of gravity in Europe.

However 'European' can have very different geographical connotations. Thus the only organisation to which every country in Europe belongs is the UN Economic Commission for Europe, in whose work the United States and Canada also participate. Some European countries belong to organisations with a common ideological basis (the Council of Europe, the Communities). The other organisations, including the OECD, have a non-European as well as European economically- or politically-based membership (Comecon, NATO).

The European Communities are the regional organisation with the greatest powers, since the Council and the Commission can adopt measures which are binding upon Member States.

To complete this brief survey of the environmental rôle of international organisations, reference must also be made to a number of international organisations or bodies of limited geographical scope and with closely specified duties. There are some forty or fifty of these, including, for example, the International Joint Commission between the US and Canada established by the 1909 Treaty on Boundary Waters. The relationship between economic growth and environmental degradation suggests they will have to play an even fuller rôle in developing transnational environmental policy in the future.

Notes

1 Except where otherwise stated, Decisions are binding on all Members and implemented by them in accordance with appropriate national procedure. One such

Decision to control polychlorinated biphenyls (PCBs), was adopted on 13 February 1973. Recommendations are mutually agreed acts which are submitted to the Members for consideration in order that they may, if they consider it opportune, provide for their implementation. Though not binding, Recommendations have a considerable impact on policy-making in Member countries. Thus, at the first meeting of the Environmental Committee at Ministerial level on 13 and 14 November 1974 ten Recommendations were adopted, covering such widely diversified fields as the control of dangerous chemical substances; analysis of the environmental consequences of significant public and private projects; noise prevention and abatement; traffic limitation and low-cost improvement of the urban environment; measures required for further air pollution control; the control of water eutrophication; strategies for controlling specific water pollutants; energy and the environment; implementation of the Polluter-Pays Principle; and principles concerning transfrontier pollution.

2 A detailed description of the OECD's work on the environment is contained in the booklet '*OECD and the Environment*', OECDg.

3 Annex to the Council Recommendation on Guiding Principles concerning the International Economic Aspects of Environmental Policies.

4 Examples include legislation by Switzerland to control the toxic substances trade, 1969; by Japan, to control chemical products, 1973; by Sweden, on products which are hazardous to man or the environment, 1973; by the United Kingdom, on health and safety at work, 1974; by Norway, on chemicals control, 1 July 1977, by France, on chemicals control, 12 July 1977; and in the United States the Toxic Substances Control act of 1976, which entered into force on 1 January 1977.

5 For example, Convention for the Protection of the Mediterranean sea against pollution (Barcelona 16 February 1976); Convention on the International Commission for the Protection of the Rhine against Pollution (Bern); Convention concerning the Protection of the Rhine against Chemical Pollution (Bonn, December 1976); Convention concerning the Protection of the Rhine against Pollution by Chlorides (Bonn, December 1976); Convention on the Protection of the marine environment of the Baltic Sea area (Helsinki, 22 March 1974); Convention on the prevention of marine pollution by dumping of wastes and other matter, (London, 29 December 1972); Convention for the prevention of marine pollution by dumping from ships and aircraft (Oslo, 15 February 1972); Convention for the prevention of marine pollution from land-based sources (Paris, 4 June 1977); Convention on the protection of the environment between Denmark, Finland, Norway and Sweden (Stockholm, 19 February 1974).

6 See the Recommendation on 'Principles concerning Transfrontier Pollution' adopted on 14 November, 1974. Recommendations are not binding upon Member countries. They are submitted for consideration in order that Member countries may, if they consider it opportune, provide for their implementation. Though not binding, Recommendations constitute a sort of moral obligation and have a considerable impact on policy-making in Member countries.

References and bibliography

Environmental Programmes of Intergovernmental Organisations, (1978 edn.) Martinus Nijhoff.

Kiss, A.C. (1977) 'La protection internationale de l'environnement', *Notes et Etudes Documentaires*, 17 Oct. 1977, la Documentation française.

OECD (a) (1972) *Problems of Environmental Economics*, Paris.

OECD (b) (1972) *Survey of Pollution Control Cost Estimates in Member Countries*. Paris.

OECD (c) (1973) *Analysis of Costs of Pollution Control*, Paris.

OECD (d) (1974) *Economic Implications of Pollution Control*, Paris.

OECD (e) (1975) *Environmental Damage Costs*, Paris.

OECD (f) (1975) *The Polluter-Pays Principle*, Paris.

OECD (g) (1976) *OECD and the Environment*, Paris.

OECD (h) (1976) *Economic Measurement of Environmental Damage*, Paris.

OECD (i) (1976) *Pollution Charges: An Assessment*, Paris.

OECD (j)) (1977) *Environment Committee: co-operative technical programme to measure the long-range transport of air pollutants*, Paris.

Appendix 1 Percentage decreases in industrial output from environmental control's effect on international competitiveness: alternative general-equilibrium predictions

Industry number and title	Polluter-pays		Subsidy-and-tax	
	Macro-orthodox	*Classical*	*Macro-orthodox*	*Classical*
Agriculture, forestry, and fisheries				
1. Livestock and livestock products	2.98	0.51	1.12	0.23
2. Other agricultural products	2.62	0.58	1.01	0.29
3. Forestry and fishery products	2.04	0.26	1.59	0.31
4. Agricultural, forestry and fishery services	1.28	0.67	1.12	1.12
Mining				
5. Iron and ferroalloy ores	1.52	0.25	0.97	0.21
6. Non-ferrous metal ores mining	1.73	0.25	1.10	0.20
7. Coal mining	2.30	0.37	1.03	0.21
8. Crude petroleum and natural gas	0.91	0.13	1.24	0.23
9. Stone and clay mining and quarrying	1.22	0.19	1.11	0.23
10. Chemical and fertiliser mineral mining	1.19	0.23	0.92	0.23
Construction				
11. New construction	1.56	1.39	1.11	1.31
12. Maintenance and repair construction	1.19	1.07	1.13	1.37

Industry number and title	Polluter-pays		Subsidy-and-tax	
	Macro-orthodox	Classical	Macro-orthodox	Classical
Manufacturing				
13. Ordnance and accessories	1.42	1.32	1.09	1.33
14. Food and kindred products	1.88	0.81	1.08	0.79
15. Tobacco manufactures	0.97	0.45	1.06	0.97
16. Broad and narrow fabrics, yarn and thread mills	1.96	1.18	1.10	0.92
17. Miscellaneous textile goods and floor coverings	2.17	1.40	1.15	0.99
18. Apparel	1.14	0.81	1.13	1.14
19. Miscellaneous fabricated textile products	1.40	0.90	1.06	0.99
20. Lumber and wood products, except containers	0.78	0.55	1.12	1.11
21. Wooden containers	1.05	0.87	1.08	1.13
22. Household furniture	1.29	1.14	1.10	1.29
23. Other furniture and fixtures	1.39	1.26	1.10	1.31
24. Paper and allied products, except containers	2.81	1.68	1.11	0.81
25. Paperboard containers and boxes	3.01	1.75	1.07	0.77
26. Printing and publishing	0.84	0.68	1.13	1.34
27. Chemicals and selected chemical products	3.75	3.25	1.03	1.09
28. Plastics and synthetic materials	4.21	3.77	1.07	1.25
29. Drugs, cleaning and toilet preparations	2.14	1.89	1.08	1.24
30. Paints and allied products	4.99	4.51	1.07	1.17
31. Petroleum refining and related industries	4.64	2.38	1.06	0.46
32. Rubber and miscellaneous plastics products	1.94	1.47	1.09	1.08
33. Leather tanning and industrial leather products	2.08	1.78	1.04	1.16
34. Footwear and other leather products	1.19	1.35	1.09	1.63
35. Glass and glass products	1.71	1.61	1.12	1.38
36. Stone and clay products	2.80	2.60	1.13	1.29
37. Primary iron and steel manufacturing	2.73	2.39	1.06	1.21
38. Primary non-ferrous metal manufacturing	3.77	1.93	1.12	0.68

Industry number and title	*Polluter-pays*		*Subsidy-and-tax*	
	Macro-orthodox	*Classical*	*Macro-orthodox*	*Classical*
39. Metal containers	1.99	1.63	1.08	1.16
40. Heating, plumbing and structural metal products	1.81	1.47	1.06	1.15
41. Stampings, screw machine products and bolts	1.66	1.38	1.08	1.20
42. Other fabricated metal products	1.68	1.39	1.08	1.20
43. Engines	2.17	2.01	1.02	1.25
44. Farm machinery and equipment	2.23	2.03	1.06	1.29
45. Construction, mining and oil field machinery	1.66	1.45	0.86	0.96
46. Materials handling machinery and equipment	2.03	1.77	1.04	1.17
47. Metalworking machinery and equipment	1.91	1.78	1.11	1.36
48. Special industry machinery	1.85	1.77	0.98	1.26
49. General industrial machinery and equipment	1.98	1.77	1.05	1.24
50. Machine shop products	1.82	1.63	1.10	1.28
51. Office, computing and accounting machines	1.32	1.50	1.28	1.79
52. Service industry machines	2.24	1.95	1.05	1.20
53. Electric industrial equipment and apparatus	1.39	1.19	1.06	1.22
54. Household appliances	1.68	1.41	1.06	1.20
55. Electric lighting and wiring equipment	1.54	1.46	1.15	1.49
56. Radio, television and communication equipment	1.14	1.01	1.09	1.26
57. Electronic components and accessories	1.39	1.34	1.16	1.48
58. Misc. electrical machinery, equipment and supplies	1.55	1.30	1.03	1.20
59. Motor vehicles and equipment	2.66	2.62	1.03	1.37
60. Aircraft and parts	1.37	1.30	1.04	1.30
61. Other transportation equipment	1.87	1.76	1.09	1.35
62. Scientific and controlling instruments	4.52	4.17	1.05	1.26
63. Optical, ophthalmic and photographic equipment	4.28	3.95	1.08	1.31
64. Miscellaneous manufacturing	2.34	2.38	1.09	1.51

Industry number and title	Polluter-pays		Subsidy-and-tax	
	Macro-orthodox	Classical	Macro-orthodox	Classical
Services, govt., dummy, and special industries				
65. Transportation and warehousing	1.01	1.18	1.17	1.71
66. Communications; except radio and TV broadcasting	0.20	0.25	1.14	1.49
67. Radio and TV broadcasting	0.49	0.51	0.96	1.47
68. Electric, gas, water and sanitary services	5.27	5.18	1.14	1.33
69. Wholesale and retail trade	0.60	0.64	1.15	1.48
70. Finance and insurance	0.46	0.50	1.15	1.48
71. Real estate and rental	0.57	0.56	1.15	1.46
72. Hotels; personal and repair services except auto	0.85	0.82	1.14	1.43
73. Business services	0.75	0.74	1.12	1.44
75. Automobile repair and services	0.98	0.93	1.14	1.41
76. Amusements	0.77	0.74	0.40	1.42
77. Medical, educ. services and non-profit organisations	0.43	0.44	1.15	1.46
78. Federal Government enterprises	0.81	0.70	1.08	1.33
79. State and local government enterprises	1.09	1.05	1.15	1.43
81. Business travel, entertainment and gifts	1.76	1.39	1.08	1.18
82. Office supplies	3.16	2.76	1.04	1.12

Notes: *Macro-orthodox* considers inter-industry linkages, substitutability of demand, multiplier effects on aggregate economic activity, exchange-rate consequences; assumes full pass-through of costs onto prices.
Classical considers inter-industry linkages, substitutability of demand, *no* impact on aggregate economic activity, *no* exchange-rate consequences; partial pass-through of costs onto prices and absorption in factor costs.

Source: John H. Mutti and J. David Richardson 'International competitive displacement from environmental control: the quantitative gains from methodological refinement', *Journal of Environmental Economics and Management* (forthcoming).

Index

Notes are indicated by n.

advertisement, commercial, 26–7
aggregation, in Forrester–Meadows
　　models, 21–3
agriculture
　　and change in land use, 51–4
　　chemicals for, 173–4, 194–5(n1)
air pollution, 174, 186
allocation, *see* cost-allocation systems,
　　intertemporal allocation, optimal re-
　　source allocation
animals, endangered species of, 174, 175,
　　180
assimilative capacity, environmental, 164,
　　168

benefits
　　net, in fishery resources, 120, 121,
　　　124–5, 128(n5), 131–5
　　social, *see* social benefits
biological (renewable) resources, 35
　　optimal use policy, 46–9, 57(n23, 26,
　　　27)
　　fish, 35, 46, 129–30
biomass, optimal, in fisheries, 112, 130–8
bionomic equilibrium, 135–6, 137–8
buy–back programme, 122–3, 124

Canada, 118–20, 121–3, 124, 194
capital, in the fishing industry, 116–17,
　　131
　　see also effort
capital assets
　　fishery resources as, 129, 133
　　in modified Forrester–Meadows model,
　　　12–16
　　scarcity as economic constraint, 10, 11
capital resources, 11, 32, 62, 63

capital theory applied to fisheries, 133–4
carcinogens, 173–4
catastrophes, 21–2
charge strategies in environmental con-
　　trol, 166–8
　　see also economic instruments
chemicals, 173–4, 190–1
coal, 10–11, 30, 41
common property resources, 6–7, 112,
　　138(n1), 164, 191
　　fisheries, 113, 114–18, 126–7, 129, 134–
　　　6
　　see also property rights
compensation criteria, 4, 94–6, 106
compensation for pollution, 188, 192
competition, primary–secondary, 76–7
competitiveness, industrial, 188
　　international, 167–81 *passim*, 197–200
　　and process pollution, 169–73
　　and product pollution, 173–7
conservation
　　criterion for material flow balance, 82–3
　　endangered species, 174, 175, 180
　　ethical approach, 79–83
　　fishery, 118–20
　　Golden Rule of Resource Conserva-
　　　tion, 134
　　optimal policy choice, 32
　　social choice and, 55(n6)
consumption
　　highest sustainable rate, 68–9
　　saturation level, 69–70, 73(n6)
cost-allocation systems, environmental,
　　166–8

damage, pollution, 23–5, 153, 173–4
　　cost curve 153, 155–6

evaluation, 24, 164–5
decision-making, 9, 20–1
demand pricing, 83–4
demand/supply analyses, 55(n5)
demographic relationships, *see* population
depletion allowance (deduction), 84–5, 86(n4)
depletion costs, 78
depletion of fish stocks, 118–20
depletion rate, optimal, 63–72, 73(n6), 88–9, 98, 106–9
and material flow, 78
developing countries, 163, 165, 178
diminishing returns, law of, 10
disarmament, international, 29
discount, social rate of, 40, 133–4, 138, 139(n11)
discount rate of future benefits, 64–6, 69–70, 73(n6), 78, 80, 86(n3)
in fishing industry, 122–3
discounts, volume, 76, 83–4
disposal costs, waste, 78, 85–6, 147–9, 153–60, 177
doom, threat of, 9
dynamic models, 10–23
policies to meet, 23–9
dynamic economic models, 9–29
value, 17, 18–19
for intertemporal choice, 32, 33
of sea fisheries, 112, 129–38
dynamic welfare optimum, 5–6, 88–109

economic instruments, 142, 144–51 *passim*
and water pollution, 153–60 *passim*
economics, definition, 62–3
effective rates of protection, 171–2
efficiency criterion for material flow balance, 79, 82–3
effort, fishing, 114–18, 123–4, 131–2, 134, 136–8
Environment Committee, OECD, 184–5, 195(n1)
Environmental Protection Agency, 173–4
equal probabilities assumption, 99, 100
equilibria, economic, 112
bionomic, 135–6

equity, *see* intergenerational equity
ethical preferences, 97–8, 99
Europe
air pollution, 186
environmental policies, 165
European Community, 165, 168, 184, 193, 194
France, 157, 186
Netherlands, 157, 186
pollution control subsidies, 168
redistributive charging schemes, 157
relocation of plant, 178–9
solid waste management, 151
UK, 66, 186
UN Economic Commission for Europe, 184, 194
exhaustible resources, *see* non-maintainable resources, non-renewable resources
export restrictions, 175
externalities, 7, 23–4, 40, 85
extraction, economies of scale in, 91–2, 95, 98
extraction costs, 34, 38
and biological resources, 47–8
of land services, 57(n28)
and non-renewable resources, 41–2, 56(n11)
and recyclable resources, 45–6, 76

'Fairness' concept, 6, 104–6
fish as a biological resource, 35, 46, 129–30
fisheries
economic theory, 112, 114–17, 129–38
rationalisation, 81, 117–27, 136, 137–8
flow resources, 35, 55(n3)
optimal use policy, 49–54, 57(n24, 26, 27)
Forrester, J.W.
model of world community, 10–12, 16–17, 19, 21
modification of model, 12–16, 21
on population growth, 19–20
France, 151, 157, 186
freight rates, and material flow balances, 75, 76, 78, 83–4

Golden Rule of Resource Conservation, 134
government intervention
  in environmental policy, 165–6, 175–6
  in fishing activity, 113, 117, 119–27, 136–8
  regulation of environmental goods, 7
  and social welfare, 106–9
  subsidies, 167–8
  and trade, 169–73
  and waste management, 144–51 *passim*, 153–60 *passim*
  see also regulations and standards

Harsanyi, J., 97–100, 103–4
harvesting costs, in fisheries, 131–2, 138–9(n8, 10)
health hazards, 173–4
housing, and change in land use, 51–4

incentives
  in buy-back system, 122
  to control pollution and depletion of exhaustible resources, 26–7, 78–83, 85, 165
  for economic substitution, 23
  harvesting quota rights, 137
  for relocation of plant, 178–9
  risk, 71
  in solid waste management, 147, 150
  and transfrontier pollution, 180
income distribution, 27–8, 82, 149
incomplete ownership, 6–7
information
  international co-operation on, 183, 191
  and OECD, 185–8, 191
  on pollution control, 168, 173, 176–7
  on waste products, 145
intergenerational equity, 5–6, 60–1, 64–9, 81–3, 90–109
  evaluation, 24, 89
  'Fairness' concept, 104–6
  and fishing licences, 124–5
  government intervention, 106–9
  Rawlsian approach, 65, 67–8, 100–2, 103–4
  utilitarian approach, 64–6, 96–100
international organisations, 183–94

European Community, 165, 168, 184, 193, 194
  multinational corporations, 172–3, 176–7, 177–9
  OECD, 165, 166, 184–94, 194–5(n1)
  UN Economic Commission for Europe, 184, 194
  UNO, 125–6, 165, 183, 194
international problems
  disarmament, 29
  environmental policy variations, 163–79, 180–1, 183–91
  fishery allocation, 81, 113–14, 125–7
  pollution, 28–9, 162, 163–81, 183, 194–5(n1)
  poverty, 27–8
intertemporal allocation of natural resources, 31–2, 60–1, 63–73, 78–80, 81–2
  analysis, 37–54
  logic, 35–7

Japan, 178, 179, 188
justice, 88, 98
  Rawlsian principles, 67

labour
  in fishing industry, 116–17, 127, 131; *see also* effort
  as a resource, 31, 32, 55(n3), 62, 63
land
  as a natural resource, 10, 11, 30–2
  in modified Forrester–Meadows model, 12–16
  optimal use policy, as flow resource, 35, 49–54, 57(n24, 26, 27)
Law of the Sea Conference, Third, 125–6
licences, in fishing industry, 118, 122–3, 124–5
literature on natural resource economics, 32–4
litter, 75–6, 84
location of plant, 177–9, 180

macroenvironment, 80–3, 86
maintainable natural resources in modified Forrester–Meadows model, 10–17 *passim*

marginal stock effect, 139(n10, 11)
market failure, 6–7, 78–9, 85, 86
in fishery resources, 117, 136
market power, 3–4, 71, 81–2
material flows, 74–86
see also intertemporal allocation of natural resources
maxim principle of justice, 67–8
maximin criterion in social welfare, 100–3, 105, 106
Meadows, D.L.
model of world community, 10–12, 16–17, 19, 21
modification of model, 12–16, 21
microeconomic theory, 80–3, 86
monopoly power, 71, 107
multinational corporations, 172–3, 176–7, 177–9, 179–80

natural resources, 30–2, 55(n1–4), 62–3
classification, 10–11, 30–2, 35
literature on economics of, 32–4
see also biological resources, flow resources, maintainable natural resources, non-maintainable natural resources, non-renewable natural resources, recyclable natural resources
Netherlands, 157, 186
non-maintainable natural resources in modified Forrester–Meadows model, 10–17 *passim*
coal, 10–11, 30, 41
non-renewable natural resources, 35
oil 31, 35, 55(n8), 71, 84
optimal use policy, 36–7, 40–2, 43–6, 56(n19)
see also flow resources

OECD, 165, 166, 184–94, 194–5(n1)
Environment Committee, 184-5, 195(n1)
Environment Directorate, 188
Group of Economic Experts, 187–8
information dissemination function, 185–8
oil, 31, 35, 55(n8), 71, 84
oligopoly, 76–7

OPEC countries, 55(n8), 71
open-access resources, 6–7, 112, 138(n1), 164, 191
fisheries, 113, 114–18, 126–7, 134–6
see also property rights
optimal depletion rate, 63–72, 73(n6), 88–9, 98, 106–9
and material flow, 78
optimal resource allocation, 3–5, 62–3
ownership
incomplete, 6–7
sole, 116, 117, 127, 129, 135–6

Pacific salmon fisheries, 118–20, 121–3, 124
paper, recycling of, 74–5
Pareto criteria
efficiency, 93–5, 106–7
optimality, 4–6, 93–5
Pigou, A.C., 65
Pigovian tax, 142
planning horizons, 34, 36, 55(n4)
Polluter Pays Principle, 148–9, 158–9, 160(n1), 166
and competitiveness, 172, 179
international variation, 167–8, 179, 189, 192, 195(n1)
pollution, environmental
air, 174, 186
control costs, 167–73, 197–200
control equipment, 177, 180
economic control, 23–5, 85–6, 142, 144–51, 153–60, 186–7
evaluation, 24, 153–7, 159, 164–5
information and OECD, 185–8
in modified Forrester–Meadows model, 10–17, 21–2
process, 169–73, 190
product, 173–7, 180, 190–1
regulation, 24–5, 142, 153–60 *passim*
water, 153–60, 171, 174
population
growth, 10–20
optimal level in fish (biomass), 112, 130–8
regulation, 28
poverty relief, 27–8
price mechanism
in aggregated models, 22–3

in free markets, 144
and intertemporal choice, 37, 71–2, 78–9
and substitution, 23–4, 71–2, 75–6

prices
and disposal costs, 78, 85
primary and secondary, 76–7, 86
primary industries, 76–7, 86
process pollution, 169–73, 190
product charges, 148–9, 150
USA legislation, 146, 151
product pollution, 173–7, 180, 190–1
property rights, 3, 138(n1)
in fisheries, 116, 126, 127, 129, 137
protection, effective rates of, 171–2

Ramsey, Frank, 65
rationality in social preferences, 97–8, 99
Rawls, J.
and maximin criterion, 100–2, 103–4
and optimal depletion rate, 65, 67–8
recyclable natural resources, 35, 43–6
recycling, 74–5
compared with biological renewal, 46
cost, 44–6
credit, 149, 150
net, 45
and price mechanism, 23–4, 75–6, 177
redistributive charging schemes, 157–60
regeneration rate, 46–7, 56(n18, 21)
regulations and standards
compared with economic instruments, 25, 26, 142, 153–6, 160
economic effects, 163, 166, 169–77, 179–80
on environmental goods, 7
international variations, 164–8, 179–80, 189–93
in Pacific salmon fisheries, 113–14, 119–27, 136–8
renewable natural resources, see biological resources
research and development, 23, 24
resource allocation, optimal, 3–5, 62–3
resource rent, 115–17, 128(n6)
allocation, 124–5
dissipation, 116, 123–4, 127
maximisation, 117–18

resource reservation, 78–9
resources, definition, 62–3
risk, 3, 4, 65–6, 71
risk-averse behaviour, 67

salmon fisheries, 118–20, 121–3, 124
satiation levels, 69–70, 73(n6)
Schaefer, M.B., 130–2
scrap
freight rates, 75, 83–4
prices, 75–7, 177
substitution with virgin materials, 76
taxes, 84–5
secondary industries, 76–7, 86
sensitivity of international economy, 172–3, 173–4
services, 21, 43–6
land, 49–54
useful resource, 40, 56(n10)
social benefits, 34, 37–40, 54, 57(n29)
and biological resources, 47–9
and flow (land) resources, 50, 51–4, 57(n28, 29)
and non-renewable resources, 41–2, 56(n11)
of pollution abatement, 153
social choice, 31–2, 164
social cost, 39, 41–2, 131
social rate of discount, 40
social welfare, 88–109 passim
sole ownership, 116, 117, 127, 129, 135–6
standard of living, 20, 28, 74
standards, see regulations and standards
static economic models, of sea fisheries, 114–17, 128(n4–6)
static welfare optimum, 3–5
subsidies, 145, 167–8
substitution effects
in aggregated models, 22–3
and price mechanism, 23–4, 71–2
virgin and scrap materials, 76
sustainable consumption rate, highest, 68–9
sustainable rent, 132–6 passim
sustainable revenue, 129, 131–3
sustainable yields, 130–2
maximum, 114–15, 117, 119–20
optimum, 117

taxes
and material flow balances, 78, 84–5
Pigovian, 142
on pollution, 25–6, 142, 145–51 *passim*,
157, 166–7; *see also* Polluter Pays
Principle
to reduce fishing effort, 137
technological innovation
in fishing industry, 120–1, 123–4
in Japan, 188
rate, 18
technological knowledge
in modified Forrester–Meadows model,
13–16
in multinational corporations, 173, 176–
7
and OECD, 185–8
and pollution control, 177
time
and intertemporal allocation of natural
resources, 32, 35–6, 39
lags and timing, 17–18, 21–2, 29(n3)
preference, 35–6, 79
trade, international barriers, 169, 171–6
passim, 180–1, 181(n5), 190
effects of environmental management,
168
and plant location, 177–9, 180
and Polluter Pays Principle, 189
and process pollution, 169–73, 190
and product pollution, 173–7, 180,
190–1
transfrontier pollution, 180, 183, 184–5,
190–3, 194–5(n1)

UK, 66, 186
uncertainty, 4, 65–6, 71
UNO (United Nations Organisation),
125–6, 194
Environment Program, 165, 183, 194
USA
Council on Environmental Quality, 171
effective rates of protection, 171–2

Environmental Protection Agency,
173–4
environmental standards, 169–71
freight rates, 83–4
and international organisations, 194
Pacific salmon fisheries, 118–20, 121,
122
paper recycling, 75
pollution control subsidies, 168
product charge legislation, 146, 148
relocation of plant, 178–9
user charges, 147–8, 149–50, 151
user cost, 38–40
and biological resources, 47–8, 56–
7(n20, 22)
and land use, 51–4, 57(n30)
and non-renewable resources, 41, 42
and recyclable resources, 43, 44–5,
56(n15)
Utilitarianism, 64–6, 69–70, 96–100, 103–
4, 106
utilitarian welfare function, 96
utility functions, 91–2, 95–8 *passim*, 105–
6, 107–9

value judgements, 88–9, 91, 93, 94, 118
virgin materials
freight rates, 83–4
prices, 75–7, 177
substitution with scrap, 76
taxes, 84–5

waste
disposal costs, 78, 85–6, 147–9, 153–60,
177
generation, 79–80
management, 142, 144–51, 153–60, 164
water pollution, 153–60, 171, 174
welfare optimum, 88–9
dynamic, 5–6, 89–109
static, 3–5